This is a Saffron

This edition published January 2018

ISBN 978-1-912496-00-6

A copy of the British Library Cataloguing in Publication Date is available from the library.

Science for the Gardener

"Don't be afraid–
a little extra Science knowledge will help you improve your garden".

By Tony Arnold MCIHort

Acknowledgements

This book has taken a year to write although the Science for the Gardener project started over three years ago when work on my website first began.

I cannot thank enough Anne Perez of Saffron Publishing for the considerable IT help, back up and support she has given me during this time, especially recently for advice on the preparation of the book manuscript and producing botanical drawings for the more technical chapters.

Support from the Royal Horticultural Society at Wisley has been steadfast for the Science for the Gardener project. This is now a PowerPoint Download for all UK teachers on the RHS Schools Campaign website as a Secondary Science Resource. I am indebted to Emma Griffiths Skills Development Officer and her team at Wisley for the constant encouragement and support for the project.

May I say it is a great honour to have the backing of the President of the Chartered Institute of Horticulture, Dr. Owen Doyle who has supported the Science for the Gardener book's aim of improving and widening the basic knowledge of all those who may be interested and involved in gardening and horticultural science.

Thanks to my patient and long suffering wife Pam for the huge amount of secretarial work she has undertaken in checking, research and an attempt to translate much of my 'too technical' writing into a more down to earth and reader friendly narrative.

A final thank you to the Hardy Plant Society for permission to use any of their superb photo library pictures and also to the contributors to the wonderful photographic library Pixabay for some unusual pictures.

Thanks especially to CGP Publications Ltd for permission to use GCSE Revise Guide images.

Foreword

"Following a rewarding career teaching science in secondary schools Tony Arnold engaged in formal horticulture education to support his passion for gardening. Tony takes the reader on a personal journey providing a "drone-eyed" view of the complex landscape that comprises the many sciences that underpin gardening. Commencing with the formation of planet Earth, we are taken on a tour touching on the highlights of plant evolution, the vast use that humans have for plants, through their cultivation and care, and onwards to modern production techniques that are useful here on Earth and for future space travel. With a relaxed conversational style Tony provides a glimpse of the complex world of plants and our relationship with them, sowing the seeds that hopefully will germinate an interest in the sciences that influence gardening."

Prof. Owen P.E. Doyle B.Agr.Sc.(Comm. Hort.), PhD (Plant Pathology), MIBioll, FCIHort, CHort.

President of Chartered Institute of Horticulture

Introduction

I have been a keen gardener for many years. Towards the end of my career as a science teacher I was lucky enough to be able to study Advanced Horticulture at Capel Manor College, Enfield and I am now a Member of the Chartered Institute of Horticulture.

Over the years I have often been asked: How can I make my garden greener? What plants should I choose and where should they be planted? How best can I grow them?

Something told me that my science knowledge and passion for gardening would sit well together and I began to think ... the seeds of an idea were sown.

Three years ago I gave a talk to my local Horticultural Society, the idea had germinated and grown into Science for the Gardener, a useful PowerPoint Presentation for gardening clubs and schools, available on the RHS Schools website.

And now Science for the Gardener, the book, which attempts to answer some of those often asked questions in a fun and light-hearted manner, with colourful pictures and diagrams along the way.

Definitely not meant to be a text book but hopefully an easy and enjoyable read which will appeal not only to gardeners of all ages but to school children, to students, indeed to anyone who may be interested in the study of horticulture and the green world around us.

With subjects ranging from the beginnings of plant life to genetic modification and climate change there is something for everyone.

I am convinced that a little science knowledge can be extremely beneficial in getting the best from our gardens and believe that as your garden begins to grow, your passion for gardening and its science will grow along with it, just as mine did.

Happy gardening, with a bit more scientific knowledge.

Tony Arnold MCIHort

Contents

Chapter 1 Plant Evolution – Algae to Angiosperms

It all started with a bang – a very big bang – nearly 14 billion years ago, or so we are told.

9 billion years later, a mere 4.6 billion years ago, planet Earth, an immensely hot magma blob, was created, steamy and foul smelling, a mixture of dust and gasses.

The beginning of time

Evolution had begun - a time line of such gigantic proportions the figures are almost impossible for us to comprehend. To think that we, the human race, and everything we see and know around us evolved from a smelly mixture of dust and gasses.

First things first, Earth needed to cool down. It took another billion or so years for this to happen and a crust to form around the planet (rather like the skin on custard).

The figures are mind boggling I know, but bear with me things do get more interesting.

As the outer layer began to cool a little, volcanoes began to appear from the depths of the still very hot planet. Volcanoes produce water vapour, lots of it, from hydrogen and oxygen. And, guess what, that made water. Good old H_2O the one scientific formula we all know. Water; we cannot have living matter without water.

The atmosphere at that time was full of very 'pongy' and noxious volcanic gases made from the elements of carbon, hydrogen, nitrogen and oxygen. I am sure you are all familiar with some of these gasses such as methane CH_4, (smelly) and ammonia NH_3 (even more smelly).

These gasses combined with the water vapour produced by the volcanoes and became what we call a primordial soup (don't suppose it tasted very good!). Just as well we humans were not around at the time; we wouldn't have been able to breathe because there was no available oxygen. It was all stuck in the soup.

Cyanobacteria, a single cell bacteria, is believed to be the very first organism to emerge from the soup, using light to make the sugars needed for growth from carbon dioxide and water, and releasing oxygen into the air. The oxygen needed to sustain human and animal life in the future. It was several thousand million years before plants as we know them evolved to be able to make the sugars necessary for growth in this way, which is known as photosynthesis.

Stromatolites

Fossilised remains of these first single cell bacterial organisms have been found around the world in Australia, Belize and the Bahamas. Scientists have named these remains Stromatolites.

Over the next 50 million years these organisms started to multiply to produce what we now know as Green Algae. Green Algae lived mostly in water - ponds, lakes and rivers or in the seas, attaching itself to the first rock formations. Not terribly pleasant, rather slimy and not very welcome on our ponds today, but they were the ancestors of the first land plants.

Fast forward to 570 million years ago! Plant evolution begins. From Algae, very simple, green and slimy, to Angiosperms, the myriad of beautiful flowering plants grown today. Five different plant groups, all evolving with differing methods of reproduction – just as well we humans didn't evolve in quite the same way, or might we, who knows what the future may have in store?

Algae, long held captive by water, made its way on to dry land, aided by gradual tidal movement, where constant bacterial and fungal action in the surrounding soil had created a medium capable of sustaining land based plants.

Cooksonia

The first simple land plants, given the name Cooksonia, had very simple branched stems, no leaves or roots but were able to live in drier conditions. They were upright and developed a vascular system for drawing up

Water and nutrients dispersing spores (a cell with a protective coating) from a sac at the tip of the stem in order to reproduce.

Cooksonia, gradually evolved into plants able to produce the leaves required for the process of photosynthesis to take place.

Before I continue, let me take you on a short journey back to the future.

Most of the very earliest plants are now extinct, along with the prehistoric creatures that roamed our planet, and can only be recognised from fossil remains. A fossil, as I am sure you know is the impression of an actual plant or animal that was alive in prehistoric times. It takes the form of stone, just like the Stromatolites, which were originally bacteria.

But let me tell you about some living fossils, three known trees in existence today which have been identified from fossilised remains as having been around since prehistoric times.

The Gingko (or Maidenhair Tree) the *Metasequoia* (Dawn Redwood) and the Wollemi Pine are the only members of their original families to have survived and as such have been dubbed 'living fossils'.

The Gingko, a survivor

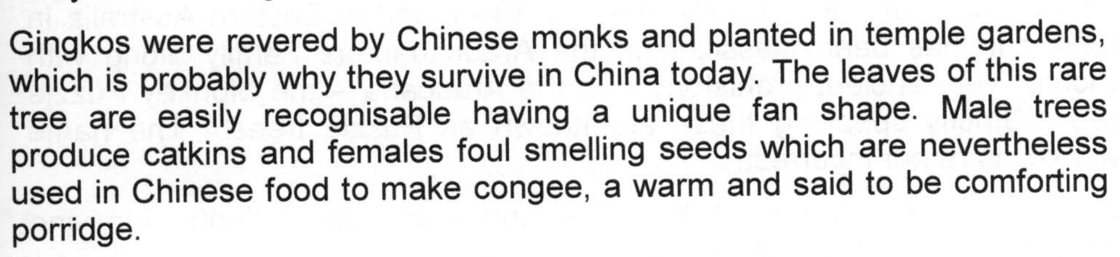

Gingkos were widespread at the time of the dinosaurs inhabiting parts of Asia and America as well as Europe.

Now they are only found growing in the wild in China but have been cultivated in many other parts of the world. They are very tough trees.

In 1945 following the atomic blast at Hiroshima, Gingkos growing just a couple of kilometres away were amongst the very few living things to survive.

Gingkos were revered by Chinese monks and planted in temple gardens, which is probably why they survive in China today. The leaves of this rare tree are easily recognisable having a unique fan shape. Male trees produce catkins and females foul smelling seeds which are nevertheless used in Chinese food to make congee, a warm and said to be comforting porridge.

Leaf extracts from the Gingko are also used in Chinese medicine and research is on-going as to whether these can be of any memory enhancing benefit, although as far as I can remember the jury is still out on this.

So all but extinct in the wild, these trees live on in plantations specifically designed for scientific research into herbal medications which are today sold in huge quantities as they are claimed to have benefits in increasing circulation, enhancing cardiovascular function and improving eye health. The Gingko contains high levels of antioxidants which may help to reduce the risk of cancer.

Of particular interest, on-going research being undertaken into the ever increasing problem of Alzheimers Disease, seeks to evaluate the benefits of Gingko in the protection of nerve cells. I can see that these trees will be cultivated for a long time to come.

Typical Redwoods

The next tree, the large Dawn Redwood (*Metasequoia*) adopted as a member of the sub-family *Taxodiaceae* and (subsequently incorporated into the Cypress Family) which was thought to have been extinct for 5 million years, was discovered in 1946 in Szechuan.

There is another prehistoric vigorous growing deciduous tree with very attractive soft and feathery pinnate leaves which can reach a height of over 100 feet. The Taxodium distichum, also known as the Swamp Cypress, is tolerant of water-logged soils where many existed in prehistoric times.

The most exciting and a very recent discovery is the third living fossil I want to mention, the Wollemi Pine. It was found in Eastern Australia in 1994 and has been classified in the Araucariaceae Family along with another very ancient prehistoric tree the Araucaria – the Monkey Puzzle tree. Strictly speaking these are known as Puzzle trees. The name Monkey is rather incorrect!

The Wollemi Pine is one of the oldest and most rare of plants, surviving since the time that dinosaurs roamed the earth.

The Wollemi pine produces both male and female cones

At the time of writing this book there were less than 100 adult trees known to exist in the wild and extensive research is being undertaken to safeguard the survival of this very important genus.

Fittingly this very special living fossil tree is named Wollemi nobilis after the wild life officer David Noble who came across the tree in a canyon in an inaccessible part of Wollemi Park not far from the Blue Mountains in Eastern Australia.

I have been lucky enough to see two Wollemi saplings, one in a small public garden in Akaroa, New Zealand and one in Bristol Botanic Gardens, so hopefully these one-time fossils will live on for generations to come. If you would like to try to grow one yourself they are available to purchase online from specialist distributors.

And now back again to prehistoric times.

Algae were the forebears of liverworts and mosses. These were non-vascular plants without roots. But haven't they done well.

Here is a bit of interesting science we need to think about. Let me explain vascular. It is very, very important. Humans, along with animals, have the equivalent of a vascular transport system based on the circulation of blood around the body. The plant vascular system is based on the circulation of water and nutrients around theplant.

It is an important and surprising fact that modern day vascular plants have retained the same system of operation as the prehistoric plants from which they have evolved!

Mosses and Liverworts came a little bit later after the algae and look like dense green mats.

You get mosses everywhere where it is damp and poorly drained. They grow on the surface of soil, rocks, plant containers and paths and are associated with damp, shady and compacted conditions. Moss can be an attractive plant to clothe bricks, stones and other hard surfaces that you might want to disguise, but it is an unwelcome sign of bad drainage in lawns and on soils in flowerbeds.

Liverwort

Liverworts, also non-vascular, growing on the soil of a plant you are about to buy, can be a sign that the plant was potted quite some time ago.

Climate change is nothing new.Our climate was once that of a tropical rainforest, hot, wet and very humid and, not for the squeamish, there were some extremely large bugs and flying insects beginning to emerge.

In the Carboniferous period around 360-290 million years ago Britain was covered in vast swamps out of which rose dense forests of the first true leaf bearing plants. Giant club mosses, tall upright plants with a forked trunk out of the top of which sprouted long grass-like leaves, grew in the swamps reaching up to 150 feet tall.

Club mosses, ferns and equisetites were the first vascular plants, sucking up the nutrients required to allow them to grow to huge proportions in a relatively short space of time. When these enormous plants died they fell back into the swamp eventually becoming the coal we know and use today

Ferns are some of the oldest surviving plants. They reproduce by spreading spores which are found on the backs of the leaves.

Some modern day ferns (such as bracken) can be very invasive as the plants have adapted to reproduce not only by spores blowing in the wind but also by growing rhizomes, a mass of creeping underground roots. Boy are they successful, but also rather wonderful with their fronds emerging from a large underground clump and gradually uncurling. Their roots are good for holding soil together in some of the more tricky planting positions such as stream banks.

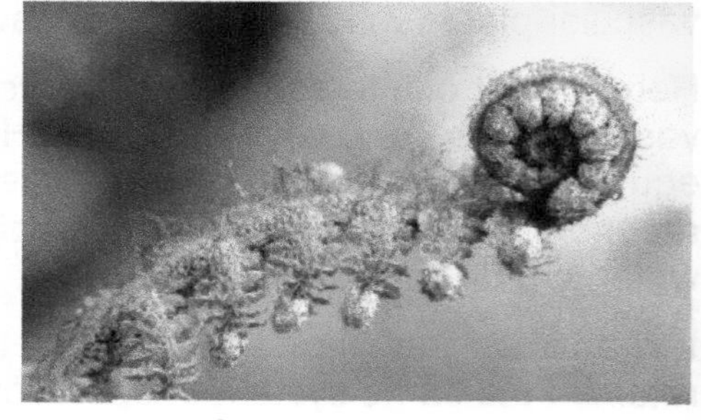

A frond uncurls

Ferns live in many different habitats and come in varying shapes and sizes. I have several different species in my garden and they are very welcome.

Equisetites, the horsetails we know today

There is another very important prehistoric plant group, the equisetites, commonly known as horsetails, which also produce spores as well as underground rhizomes similar to some species of ferns.

Equisetites in prehistoric times grew to be over 100 feet tall, but have evolved to become the much smaller modern day plant, the equisetum. They can be very attractive – they look great in a tall glass vase - but unfortunately again can be very invasive so be careful if and where you plant them. They have an uncanny habit of not waiting to be planted, but just turning up uninvited as unwelcome weeds all around the world.

Cycads are an interesting group of sub-tropical plants which lived over 200 million years ago, even before the dinosaurs, so we are led to understand. Cycads have very wide woody trunks with an impressive crown of evergreen pinnate leaves and can be several metres tall, or just a few centimetres. They are very attractive plants which are related to palm trees and tree ferns but are not suitable for growing outside in cooler temperate climates.

Cycad cones

Unlike ferns, cycads bear both male, pollen producing, and female, seed bearing, cones, the size and shape of which are very impressive, particularly the male cones! They have a pollinator which is a special species of beetle. These plants grow very slowly and can live a very long time – up to 1,000 years.

So cycads evolved to produce seed and have adapted to live in many and varied conditions. They can be found in swamp and boggy conditions, in a harsh dry desert or a wet rainforest climate, in full shade or full sun, although they are an endangered species and today there are just 250 species compared to over 300,000 in prehistoric times. I find them fascinating survivors that have adapted very well to their surroundings, having had plenty of time to do so!

Starting their life in the same time period as Cycads are the Conifers. They are the largest group of living Gymnosperms (which means they bear open, unprotected seeds). They are non-flowering plants and mostly evergreen.

Some conifers have broad leaves but most have simple needle- like leaves like Pines or scale leaves as with the Cypress - *Cupressus* Family - and then there are some with very soft feather shaped leaves as with the Swamp Cypress Family.

Conifers prefer a colder climate and they can often grow to huge sizes. One well known example, a Californian Redwood (*Sequoia*), is not only over 70 metres high, but is nearly 10 metres in circumference.

Now for the biggest changes of all and those plants which evolved to become the Angiosperms (the flowering plants) the most important group. They produce seeds enclosed in an ovary and give us the huge variety of flowering plants, trees and shrubs we know and love today.

Evolution in plant reproduction started with the simple spore, open to all elements and reliant on wind to aid dispersal. Then came those 'naked seeds' produced by non-flowering conifers with just the protection of a cone 'overcoat', and finally the Angiosperms, the flowering plants whose seeds are multi-celled fertilised ovules held in an ovary along with their own food supply.

A diverse bunch the Angiosperms.

Chapter 2 Angiosperms – the flowering plants

Now let me explain some of the science behind Angiosperms, the largest group of plants in the world, making their way onto the world's stage approximately 130 million years ago - which in fact is regarded as very late in evolutionary time - and boy did they revolutionise the plant world!

Angiosperm – a big name for a small weed

Dandelions are Angiosperms, so are orchids, grasses and horse chestnut trees, so you can see there is an enormous diversity in this group.

The name Angiosperm comes from the Greek, angeion meaning vessel and sperma meaning seed.

Like the Gymnosperms (the Conifers) Angiosperms have seeds, but here the similarity ends. The seeds of Angiosperms have the shelter and protection of an ovary. Biologists and Botanists think that it is the small size of the male pollen combined with the female protective ovary in flowering plants which has made the production of Angiosperm seeds and the resultant plants so successful. This is the reason why this group of plants has been able to diversify so well, producing many different species and hybrids.

The most important evolutionary change is the ease with which seeds are produced and the fact that Angiosperms can hybridise, which means they can reproduce between similar species.

But what is it that makes Angiosperms so special, setting them out above the first known and more basic plants? Simply, it's the flowers. Angiosperms are the 'show-offs' of the plant world having evolved to produce flowers, often very colourful or with an enticing scent. Well we all know what a flower is, but what is its purpose other than to look pretty or smell nice?

The flower is the seed bearing part of a plant. Seeds that will produce more flowers, often in abundance, and not just those grown for show, but vitally the huge range of grains, pulses, fruits and vegetables required to feed not only the human race but all living creatures here on Earth.

Although most Angiosperms have both male and female parts (the sex organs), they usually need an outsider to help them get it together. The main aim of that colourful, often sweet scented flower is to attract that helper – a pollinator.

Where would we be without bees?

A pollinator is required to transfer the male pollen to the ovules in the female organ, the ovary. Pollinators are usually bees, flies and other insects such as beetles but one rather unpleasant and unusual pollinator is the slug (yuk!) which will pollinate Aspidistra plants. Aspidistras are rather old-fashioned house plants I personally think their attraction to slugs is one very good reason not to grow them outside!

Some flowering plants can self-pollinate and some can be pollinated in water including streams and rivers. So these are what might be called the natural pollinators, but of course, then there are us humans, taking little bits of pollen off one plant and adding it to the next. Plantsmen are able to create designer plants with especially attractive flowers using tiny brushes to take the pollen from one plant and brushing it onto another. These 'man-made' species are called cultivars and what a variety of different cultivars we have helped to create. Cultivars produce more prolific and eye pleasing ornamentals for the flower markets as well as, and of even greater importance, better quality, better tasting and longer growing fruit and vegetables.

Plant reproduction is a fascinating subject which will be dealt with later in this book. Unlike humans, plants have more than one way of reproducing.

Angiosperms are divided into two groups, monocotyledons (monocots) and dicotyledons (dicots). Monocots encompass plants such as bulbs and grasses (both lawn, and those we eat, such as wheat and barley). There are also a few edible monocot trees and shrubs such as the Palm and *Musa* (the banana shrub).

Monocots, as the name would suggest, have only one seed leaf, which emerges from the germinating seed. Germination is simply the term used to describe the first sprouting seed leaf or leaves to appear once a plant has been fertilised.

A typical monocot leaf

Nearly all plants that form true bulbs (which are the organs that store the glucose-sugars needed to feed the plant) are monocots and these include ornamentals such as crocus and tulips as well as many edibles such as onions and garlic.

The leaves of monocots are 'strap-like' and petals are borne in multiples of three. The vascular vessels are in more uniform bundles than dicots and they have fibrous roots.

A typical dicot leaf

Dicots, the second and largest group, have two seed leaves which grow out of the germinating seed and most flowering plants, shrubs and trees are dicots. These plants mostly have broad leaves and petals are borne in multiples of four or five. They often have tap roots.

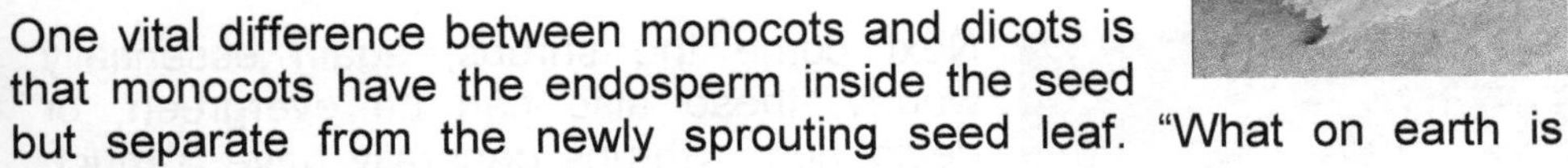

One vital difference between monocots and dicots is that monocots have the endosperm inside the seed but separate from the newly sprouting seed leaf. "What on earth is endosperm"? I hear you say.

Endosperm is one of the three vital parts of the seed that feeds the emerging seed leaves and is usually starch or glucose. It can also contain protein and vegetable oil. With dicots the endosperm is held within the leaves and is nearly all used up by the growing plant - that is until photosynthesis takes over.

Seeds, germ, endosperm and bran and the fruit and vegetables they become will be covered in more detail in the next chapter. Meanwhile, let's take a more detailed look at Angiosperms. There are over 300,000 species of Angiosperms which can be split into several different categories, all producing flowers, although not always very large or showy.

I'll start by setting out the divisions so it's less bewildering. There are trees, shrubs, herbaceous perennials, biennials, annuals and also ephemerals which are slightly less known but nevertheless important.

Firstly the trees, excluding Conifers of course which do not bear flowers and are therefore not Angiosperms. Trees are by definition woody, their trunks having evolved over millions of years from those first very spindly stems of the Cooksonia, the first land plant. They can have many branches or just a few, be weeping, bending towards water, or upright reaching for the sky looking down up on us.

Trees vary hugely in size, shape and age; they can be very long-lived. Some, such as the oak, can be hundreds of years old. Imagine what they have seen - if only they could speak, but perhaps that is taking evolution too far.

Flowering trees are usually broadleaved and often deciduous, hormones causing them to lose their leaves in winter months as the temperature drops. It's not only teenagers that suffer the effects of hormones!

Some semi broad leaved trees such as *Quercus ilex* (Holm Oak) are evergreen and very prolific growers too as a result of not shedding their leaves.

Hydrangea, blue for acid soil

Next come the shrubs, again essentially woody, these also can be evergreen, or deciduous. Shrubs generally have a much shorter main stem than trees with multiple stems growing from the base (similar to many Conifers).

Shrubs, like trees, do not have a set life cycle. They can be short-lived but given the right soil and growing conditions can last a very long time. This is where you need to be in control, shrubs do need to be kept in check, or else you will find they rather take over your garden and become very large, crowding out light and air to the detriment of some of the smaller and perhaps more delicate plants.

The remaining groups that fall under the Angiosperm heading have differing life cycles, and these can be rather confusing. Let me explain.

Herbaceous plants have no persistent or permanent woody stem above ground. Herbaceous perennials live for more than two years, although often much longer. During the winter months these plants disappear out of sight, but come spring lo and behold something amazing has been happening underground and we see those first welcome green shoots of spring appearing.

Life cycles

Herbaceous Plants do not have a woody structure. Stems and leaves normally die back in winter. They can be divided into four groups

Annuals - Complete life cycle in one year then die

Biennials - Life cycle takes two years. Leaves and stems grow in first year, flowers in second.

Perennials - Live for more than two years and survive winters outdoors

Bulbs, corms and tubers - all store food to enable them to survive a dormant period.

Different species of perennials have developed different storage organs, be they bulbs, corms, tubers or rhizomes or underground swollen stems or roots, all are designed to contain the energy necessary to feed and keep the plant alive. Energy is made up of sugars and kept in the form of carbohydrate glucose or stored as starch. These sugars keep the organ's cell tissue safe from being destroyed by frozen water as the freezing point of sugar falls far below that of water.

Ginger is a rhizome

Bulbs and some corms are not nearly as tough as some hardy tubers and rhizomes and roots. There are of course always exceptions. Dahlia tubers and many tulips for example, cannot survive in colder, wetter areas and must be removed before winter.

Similarly robust bulbs such as Daffodils (*Narcissus*) that may well survive low winter temperatures may still perish in very wet conditions, so well drained drier soil with added grit is required - be warned

Half-hardy perennials, also known as tender perennials, give up when the frosts arrive (I know how they feel) - you will need to bring them indoors if you have the space. I kept a half-hardy *Pelargonium* (known by many as the *Geranium*) alive for several years, bringing it into the conservatory as winter approached and taking it back out again the following year. Then one year I left it outside and sadly it was no more.

Tender (half-hardy) plants will not usually survive below 15 degrees, although they will be fine in tropical areas and can be found happily growing wild in abundance throughout the year in these warmer climes.

Biennials have a two year life cycle. During the first year they grow simple roots, stems and leaves but they do not normally flower until the second year. Wallflowers (*Erysimum*) are a well- known example of a biennial. Once they have flowered in year two they should be removed and replaced. Some prolific seeders such as the biennial Foxglove (*Digitalis*) may shed seed in year one, but only of course if flowers have already been produced.

There are, of course, exceptions – there always are with plants!! If a period of extreme cold for five or six weeks (which we call vernalisation) is followed by a period of exceptionally warm weather, then biennials may be tricked into flowering the first year.

From a strictly botanical view biennials are often regarded as short lived perennials: As they are short lived they readily self-seed after plentiful flowering. These flowers are very popular with pollinators.

Forced vernalisation can be induced by growers by refrigeration. This is a vital process enabling growers to obtain biennial seeds without having to wait for two years. Biennial vegetables such as carrots, cauliflower, sprouts and onions will normally be produced in year one especially if grown in a cold northern climate, but the flowering that follows, necessary to produce the seeds, may well take two years. Interestingly in very warm climates the only way growers can produce cold climate vegetables is by having large scale refrigeration induced vernalisation as part of the growing production process

Annuals have the shortest life cycle, just one year and then they die. But, happily that's not the end of them thanks to genetic adaptation. Annuals along with biennials are very prolific in setting seed to ensure the species continues and they are usually very successful. Think, for example, of those classics, the annual daises (*Bellis perennis*) reappearing every year, happily growing at the side of a busy road or in our parks and open spaces, where children pass the time making daisy chains for a loved one.

There are annual vegetables such as squashes and tomatoes (although strictly, of course, a tomato is a fruit!) Annuals also come as hardy or half hardy, and as you would expect, half hardy annuals will not withstand northern cold and frosty temperatures. Do not be tempted to put out those very attractive bedding plants that appear in garden centres at the beginning of spring until all chance of frost has passed, the plants will not thrive and become an easy and tasty target for the emerging slugs and snails.

Spring, a time for Bluebells

Ephemeral is the name given to those opportunistic plants which can develop quickly if conditions are right. They are fast growing, and can have more than one (usually short) flowering cycle a season, setting seed before once more disappearing from sight and becoming dormant until conditions are right for their re-emergence once more.

In woodland, spring ephemerals such as the Bluebell (*Hyacinthoides non-scripta*) and the *Anemone nemorosa* take advantage of the increasing light levels to flower, but then disappear once the trees form their shady canopy.

Then there are the weedy ephemerals which will spring into life and reproduce at great speed where there has been human disturbance. So be careful when you dig that new flower bed or introduce some new top soil you may just get more than you bargained for. For example each Fat Hen (*Chenopodium album*) plant can produce 70,000 seeds. Similarly Chickweed and Annual Meadow Grass are also very prolific and fast growing.

Never underestimate nature's talents for adapting plants to take the very best advantage of the surrounding conditions.

A comfortable lodging place

Epiphytes are unusual in that they grow on another plant such as a tree but are not part of the supporting plant's structure as a parasitic plant needs to be. In simple terms it is independent as it obtains its nutrients from air, rain and other plant surrounding detritus

There is a whole family called *Bromeliaceae*, the Bromeliad species that grow as Epiphytes, one example being the Pineapple (*Annanas comosus*) which is a rare edible bromeliad fruit first discovered in South America. Pineapple flowers and many Bromeliads are popular as houseplants in temperate climates. These plants do have to be admired as clever adaptors with their colourful flowers and showy bracts.

We, as gardeners mostly grow flowers for our pleasure for their beautiful scents, colours and forms that brighten our day. We all have our favourites, some that will grow better than others.

Fashions in flowers change – what is in this year? Are Roses in vogue? Will Dahlias make a come-back? The choice is yours. And what a choice there is! I can never leave a garden centre without the one extra plant that I know I will find a space for somewhere in my garden – if only those plants that I bought last year had not grown quite so large!

Gardens, of course, come in many shapes and sizes. There are designer gardens, trees and plants individually chosen for their particular size, colour or shape.

Gardens with formal borders, wild flower meadow gardens seeding themselves in the long grass, rose gardens (so many different colours and varieties) front gardens and of course the compact urban courtyard garden which can nevertheless come alive with a variety of colourful pots and planters.

Then there are those gardens we can visit - take inspiration from, the UK National Trust Gardens open to the public, the Botanical Gardens often linked to universities, the iconic RHS gardens and flower shows and of course there are those keen home based gardeners, happy to show off their hard work to others in return for a nominal donation to charity under the National Gardens Scheme in the UK.

Many European countries especially France, Italy and Germany have a wonderful range of gardens well worth a visit if you are in the area. In the USA there is a 'Top ten best public garden category' that includes the Botanical Gardens in Richmond, Virginia and the well-known Missouri Botanical Gardens in St. Louis, but do check through the other eight if you are lucky enough to be touring in the USA.

Some designers have created gardens to cover the tops and sides of buildings, for insulation purposes as well as visual impact. And then there are those planting schemes being developed to decorate the outside of some of the more affluent commercial buildings around the world. There is a well advanced plan to build a garden bridge over the central London River Thames – although money (or should I say lack of it) would, as I write this, seem to be an issue!

Before moving on from this fascinating range of flowering plants, perhaps I should mention just a few of the more unusual ones to be found around the world.

The Pitcher plant waits for lunch to arrive

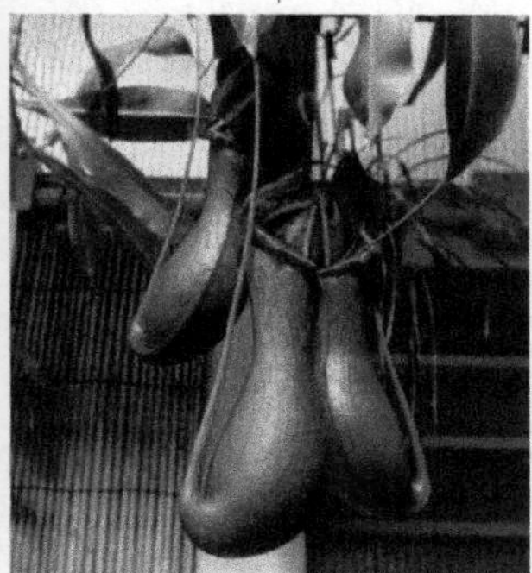

The carnivorous (meat eating plants) capturing and devouring that inquisitive fly. The two best known carnivorous plants are the Venus Flytrap (*Dionaceaemuscipulata*) from eastern USA and the Pitcher Plant (*Nepanthaceae* and *Saracemiaceae* Family species) from the South American rain forests. Pitcher Plants are also found in north USA.

Baobab, the world's largest succulent. Water is stored in the trunk

The Baobab plant, a member of the mallow family, found in Australia appears to have its roots growing upwards towards the sky, hence its nickname, the Upside Down Plant (*Adansonia digitata*).

There are two plants that give off the smell of a rotting corpse – not for the squeamish! One is the *Amorphophallus*, (*Araceae titanus*) which in fact is a fungus and the largest unbranched flower in existence growing in the rain forest of Sumatra. The second of these, the Corpse Flower (*Rafflesia arnoldii*) found in SE Asia, is slightly different – it is a parasitic plant, not a fungus, relying on obtaining nutrients from other plants. It has no leaves or roots and normally hides within its host plant. For anyone lucky enough to be visiting Kew Gardens, there is a specimen of the Corpse Flower in the tropical glass house. If you just follow your nose – there is an awful stink!

Amazingly, there is a sensitive plant, the *Mimosa pudica*, found widely in the tropics and also making an appearance in Kew Gardens. Touch its leaves and they fold up which is thought to be a defence against

herbivores. This occurs as hormone stimulation causes water to be withdrawn from the vascularsystem.

Another plant, the Lithop (*Lithopruscioru*m) from southern Africa, uses camouflage as a defenceIt looks just like the stones that surround it.

Mimosa pudica

Looks like a huge dead spider to me

On a light-hearted note one rather unfortunate plant, Tree Tumbo (*Welwitschia mirabilis*)of the *Welwitschiaceae* family found in South-West Africa is famed as being the ugliest plant – but, of course, 'ugliness is in the eye of the beholder'.

Those of you who don't like the dentist (and who really does?) avoid looking at the bleeding tooth plant (*Hydnellum peckii*) which looks like a fungus covered in spots of blood. Urghh it looks horrible. It's found in tropical Indonesia – Sumatra.

If you are a sportsperson, especially in North America, you might be interested in the baseball plant, *Euphorbia obesa*, found in deserts globally, does look rather like a baseball with stitches.

Frankincense (*Boswellia sacra*) of the *Burseraceae* Family produces a tree resin which is fragrant and can be used as an anti-inflammatory and also of course for the well-known religious ceremonies.

Orchids are a huge species, one of the largest, growing in diverse conditions from small northern meadows to the hottest tropics found in Singapore. They are very popular houseplants for colour, and sometimes for fragrance, requiring little maintenance throughout the year.

The Giant Water Lily (*Victoria amazonica*) of the *Nymphaceae* Family is a favourite in glass houses. A good example can be found at University and Research Botanic Gardens Kew.

The huge circular leaves look big enough to stand on. I like to imagine some nymphs playing hide-and-seek on them – perhaps that's where the family name came from.

Too big for my pond!

These Giant Water Lilies were one of the earliest flowering plants and can be found naturally in areas of South America such as Argentina and Paraguay. Their scent attracts beetles as pollinators but they have to be careful as the floral cup is protected by spines and the bud really heats up to release its alluring perfume. The beetle is trapped for a night after pollinating the female flower which changes colour and then sinks below the water level to set seed releasing the beetle to live another day - mission accomplished by both beetle and flower.

Mangrove plants of the *Rhizophoraceae* Family are unusual in that they act as giant filtration systems at junctions of sea and fresh water in estuaries. They rely on tidal movements to supply nutrients in soil sediments for growth.

I thought I would share with you a couple of interesting oddities I have come across in my research. Firstly, two of the longest surviving individual trees have both been discovered in the White Mountains of California. The first a Great Basin Bristlecone Pine (*Pinus longaeva*) and given the name Methuselah was thought to be the oldest at 4,845 years but was subsequently demoted by another tree of the same species said to be 5,062 years old. I wonder how much longer they can survive.

Finally, something rather peculiar and again from California! The horticulturist and tree sculptor Axel Erlandson who lived in California for most of his life, although Swedish by birth, managed to shape trees into some very unusual forms. Trees which appeared to have legs, trees that appeared to be growing not from the ground, but from the top of an archway, trees that appeared to be woven like baskets. Take a look on line – you will see what I mean. Clever, but I am not in favour of creating these weird looking structures – give me what nature provided any day.
But enough - what are Angiosperms really about, and why can't we live without them?

Chapter 3 Feeding the world

Over thousands of years man has developed the knowledge and ability to use plants to his own and best advantage.

An interesting book was published in 1883 by a botanist called De Candolle which charted the history of plants used for food over the previous 4000 years. That book was written during the reign of the great Queen Victoria - imagine if De Candolle were alive today – how amazed he would be to see the huge variety of edible plants now available to us. I wonder whether or not Queen Victoria would have been amused by the variety on offer.

Edible plants are the incredible living producers of many of our day to day food necessities. Where would we be without them? We owe our existence to the plants that grow and feed not only us but also our animals, birds and even the insects and bugs which play their vital part in the great food chain.

How many of us stop to think what part of the plant it is that we are eating, or where it has come from, other than the local supermarket? How far has it travelled to get to us, how many people have been involved in its production and can we continue to grow the amounts needed to feed an ever increasing population.

Many people are interested in various food ingredients from vitamins, minerals, amino acids to healthy fats and anti-oxidants.

What's on the label?

Check out the packaging before you check out from the supermarket, you may well be surprised what the product actually contains.

Growing techniques in horticulture are constantly being researched from the time when glass houses were found to speed up crop growth to more modern day poly tunnels discovered when a Spanish grape grower found double sized grapes under a piece of waste plastic that had chanced to fall upon his vine.

Chemical fertilisers had huge effects on increasing crop productivity in the 20th century but now concern over soil erosion along with greenhouse global warming from carbon emissions has encouraged a new mode of growing called organic growing. This is an attractive eco-friendly strategy based on renewing the organic part of the soil with suitable plant vegetative material – waste not, want not - recycling on a large scale. Artificial chemical fertilisers are not used, nor are chemical products used to help eliminate pests and diseases. Instead, certain natural plant extracts are used which are regulated by the Soil Association in the UK and the equivalent body in other countries.

Serious concerns have been raised about the declining numbers of bees and insects for pollination. It would appear that many bees have been unable to find the way back to the hive. It is thought that this could be caused by the huge amount of insecticides still in use today but there is obviously still a great deal of research needed to see if this is the case and what can be done to overcome the problem. Biodegradable pyrethrins derived from chrysanthemums are being studied as they are believed to be less toxic than insecticides.

The idea behind Genetically Modified foods is to alter the natural genetic make-up of a plant to allow better growing and protection of crops particularly in tough growing regions which may be subject to decimation by pests and diseases. Are you in favour or against? Discussed further in Chapter 12.

Scientific, horticultural and agricultural advances do of course play a huge and important part in ensuring the increasing demand for food around the world is met.

Plants for food are grown in very different climatic zones, on vines, on trees, on shrubs, above ground and below. Some plants like it hot! Others need a cold spell before growth takes place. Some need plenty of water and others have adapted to near drought conditions.

So let's take a look at some of the items in our supermarket trolley, and discover what it is we are actually eating and where in the world has it come from.

Grain crops are perhaps the most important of all and can be divided into five separate groups, cereals, pseudo cereals, pulses, whole grains and oil seeds.

We need a lot of wheat

Originally grown in the Middle East, grain crops are now grown world-wide. The best known are wheat, oats, barley, rye and corn, all monocots from the humble grass family (*Poaceae*). Once the seeds of these grasses have been harvested and dried they are known as cereals.

Cereal is any grass cultivated as a grain crop and comprises mainly carbohydrates, fats, oils and protein, as well as vitamins.

The great advantage of cereals is that they are durable, can be stored in grain silos, and are easily transported.

A huge proportion of the world's population relies on the cereal grain crops of wheat and rice grasses to form the basis of their diet.

Wheat grows mainly in the temperate areas of the world and production for 2017/18 is predicted to be around 735 million metric tons. That's a lot of wheat!

The UK grows about 85% of the wheat needed to feed its population. Wheat is ground to produce the flour used in bread, cakes, biscuits and pastries. It is also used to make pasta – contrary to the 1950's April Fool's Day spoof on UK TV, spaghetti does not grow on trees! Not sure they would have fallen for it in Italy!

But do you know what part of the plant is used? The grain seeds along with the endosperm (that word again) produced by the grass to sustain its growth are harvested and ground into flour. White flour uses only the endosperm; whole wheat, as the name would suggest, uses all the grain.

The grain seed contains the germ, the reproductive embryo, which when germinated will grow into a new plant. The endosperm feeds the seed leaf and the outer seed covering, called the husk or pericarp, produces the bran. So those three vital components together make up the whole grain and processed separately give us wheat germ from, would you believe, the germ, flour from the endosperm and bran from the outer husk. These husks are then milled to a powder form of the bran

Sorghum, a monocot of the *Poaceae* cereals family, is a major grass grain crop grown mainly in Australia but also on the African and Asian continents. It can be used as a cereal grain but is usually grown for animal fodder. There are many more grains of high importance for fodder for livestock such as Maize (Corn), Oats, Barley and Millet.

The old way

Rice is the staple food for over half of the world's population. It is the grain seed of swamp grass and requires a good deal of water to grow successfully. In many Asian countries rice is still planted manually in paddy fields, which are then flooded to deter weeds and vermin but things have moved forward somewhat in the USA where rice seed is often distributed by helicopter with water levels being strictly controlled by computer based irrigation systems.

True cereal grains come from grasses, but then there are what are known as pseudo cereals, which are grains and seeds from other non-grass families of mostly dicot plants.

Strange looking but versatile

Quinoa (from the *Amaranthaceae* Family of 'Love Lies Bleeding' fame) is an up-and-coming gluten free grain (not a wheat cereal) used as a rice substitute. Buckwheat of the *Polygonaceae* Family (again, not a wheat cereal and in fact related to Japanese Knotweed!) is another example of a gluten free grain, popular in Russia and Canada as it is very hardy and will grow in very harsh cold climates.

Breadfruit from a tree belonging to the Mulberry Family is perhaps a little known but rather diverse food product – it can be used as a base for both savoury and (when soft and overripe) sweet dishes. It can also be processed into gluten-free flour. The fruits which have a rather lumpy green flesh and the texture of potatoes are rich in vitamins and minerals, gluten-free carbohydrate and protein. A fruit can weigh up to 3kg and although they grow on trees can be cooked as potatoes and other underground tubers.

I have not tasted breadfruit myself, but I have read that some varieties actually smell like fresh bread which in my view is an absolute bonus. Although the actual taste is rather bland, as is I suppose that of the potato, seasoned well it can be added to salads, soups, almost any dish to make it go a little further.

Not just for food – interestingly the male breadfruit can also be used as an insect repellent and the sap has been used for water-proofing and chewing gum!

Acacia (commonly known as Wattle) is the largest genus of vascular plants in Australia.

A useful source of minerals and calories

Wattle is the national floral symbol of Australia and its seed (a pseudo cereal) was widely used by Aboriginals as they traversed the very difficult growing conditions found in Northern Australia. The seeds were crushed to make flour which was then added to water and made into damper (a simple flatbread) cooked on hot coals. Perhaps the damper was used to make a witchetty grub (moth larvae) sandwich, another Aboriginal delicacy!

Crushed Wattle seeds are also used as a thickening agent in sauces and casseroles and as flavouring for ice cream. Dry-roasted the seeds give off an aroma of coffee, and they can be used as both a beverage and as an addition to desserts

Wattle seeds contain a high concentration of minerals and may be of interest to the rising number of diabetics as they have a low glycemic index providing a source of sugars that do not cause peaks in blood glucose levels.

Most pulses come from the *Legume* Family which produces its seeds in pods. The dried seeds are known as pulses. Dried peas, beans, chickpeas and lentils all come under this heading and are rich in proteins, carbohydrates and essential amino acids.

Lentils are one of the oldest domesticated pulse crops and are most often used in spicy soups and stews being a good meat substitute.

Nuts are a good source of protein and reports have suggested that some varieties may play a part in lowering levels of 'bad' cholesterol. They are rich in calories producing energy, contain antioxidants, omega 3 fatty acids, vitamins and trace minerals.

Nuts mostly grow on trees and are fruits which have developed a hard shell to protect the edible kernel within. These shells will not open naturally and need the help of a hungry squirrel or human with some crackers to release the tasty innards. Chestnuts and hazelnuts as well as acorns are examples of true nuts. The trees bearing these all grow in temperate zones, where there are no extremes of hot or cold.

Peanuts are attached to roots

Peanuts are of course an exception, the nuts of this leguminous plant (in the same family as peas and beans) grow underground. Peanuts grow in many countries, but China is the world's largest producer followed by India and the USA. Jimmy Carter the 39th President of the United States of America was a one-time peanut farmer. Beware peanuts though – as with many nuts they can, to an unfortunate few, cause a severe and potentially fatal allergic reaction!

A fruit is the ripened ovary containing the seeds of a female plant. See Reproduction chapter. It is normally regarded as the edible fleshy part covering the seed, varying greatly both in taste and texture, but a hard nut is also a type of fruit

Apples (which are in fact fruit pomes) grow happily in cooler temperate climes, are best eaten when they are crisp, can be sweet or sharp, with varieties grown for cooking as well as eating straight from the tree.

By the way, in case you were wondering, a pome is a fruit that doesn't develop from the ovary, but another part of the flower or plant. Pears, Medlars (a small brown apple-like fruit only edible once it has started to decay – yuk) and Quinces also come under this category. You may find them described as Pseudocarps rather than the true fruit which is the normal fruit grown from the ovary.

Oranges and grapefruit of the citric *Rutaceae* family are softer inside, juicy and sometimes quite acidic (grapefruit, especially might need just a little sugar) needing the hotter weather of places like Brazil, Florida and California to encourage the production of fruits with their tasty citric juices.

Pineapples, tropical fruit natives of Brazil and Paraguay, spread throughout South America and on to the Caribbean and Mexico. They grow from a leafy plant out of the ground, not from a tree as often thought, and require a warm climate to thrive. Pineapples belong to the *Bromeliad* Family and are epiphytic which means that they can grow on another plant, using the host as a support, but still deriving nutrients and moisture from the air.

A good sized bunch

Bananas (Musa) are found growing on large tree-like monocot shrubs, technically described as herbaceous perennials. Strictly speaking in botanical language, a banana is a berry as it has multiple, very tiny, seeds. Bananas are sometimes used to make beer and wine! I didn't know that, did you?

Berries are small soft fruits containing seeds. Strawberries (technically not berries as the seeds grow on the outside) growing on runners across the soil's surface, gooseberries growing on bushes – but mind the prickles. Black, red and white currants and delicious raspberries growing on shrubs that need canes for support - all growing happily in the UK. Our climate's not that bad after all!

Requiring an acid soil, *Vaccinium* shrubs in the *Ericaceae* Family are very attractive with a range of colourful berries. The commercially cultivated varieties include cranberries (a Christmas treat with the turkey), blueberries, bilberries and lingonberries. Used in both sweet and savoury recipes these berries contain a high proportion of antioxidants and vitamin-packed flavonoids (the plant chemical responsible for the colourful berries) – so, not only do they taste good, they are good for you.

A sticky business

Date palms, *Phoenix dactylafera*, large growing monocot trees, thought to have originated in the Middle East, produce big bunches of sweet, sticky fruits, a Christmas treat for many.

These palms have separate male and female trees.

Pollination is achieved either by the use of a wind machine which is used to blow the male pollen towards the female plant, or by employing skilled human pollinators who, with the aid of special climbing apparatus, scale the female trees to introduce the male pollen to the female flower.

Southern Europe has the best growing conditions for olives, *Oleaeuropeae*. Black or green varieties used in salads, for nibbles with drinks but mainly as olive oil, drizzle it on your salad and dream of those sunny days spent lazing in the sunshine. And just to keep us in the right mood, we must not forget grapes.

Red for me please

Grapes for our fruit bowl, grapes with cheese and grapes for wine, and then more wine

By definition a vegetable is the edible part of a plant, although fruits, nuts and cereals which I have already talked about are not treated as vegetables or fruits from a commercial point of view. To complete (or confuse) the picture many important vegetables such as avocado, beans, pea pods, corn, squashes, nuts and grains are also fruits.

We eat leaves, stems, swollen stems, roots, tubers and corms (all of which are the organs that store a plant's own food). I am not going to give you a huge list of vegetables. I am sure you all have your own favourites ... but to give you an example of what part of the plant it is you are eating take a large pan and add the following

Potatoes (swollen stem tubers) .. carrots (tap roots) .. onions (bulbs, perennial storage organs) .. turnips and swedes (round root tubers).

Perhaps some cabbage leaves, and a few peas (those podded seeds that have escaped being dried). Add some barley (a cereal crop) and you have the basis of a very tasty and highly nourishing soup from a wide range of vegetable plants.

Cellulose and lignin fibre based in the vegetables we eat cannot be broken down by our human digestive system, so passes through our bodies as fibre and roughage which is why vegetables are very highly recommended for our digestive wellbeing.

Non-meat eating animals such as horses, cattle and sheep have the necessary bacteria and other microorganisms to break down the chewed cellulose called the 'cud'.

Plants can convert unused carbohydrate to starch, but the human body sadly cannot store unused or undigested starch and converts it to fats instead. There is varying opinion on where the fat is stored.

For our summer salads, we eat lettuce leaves, cucumber and tomatoes (these latter two are in fact strictly fruits as the seeds are grown within the fleshy surrounding) beetroot, a tap root beet crop, celery a crispy stem stalk, add a little herbaceous cress, a few olives for texture and what a healthy and colourful plate of food you have.

So many different plants grown to put everyday food on our plates. Although there are many different varieties of vegetables grown in the UK, East Anglia having particularly good growing conditions, people have become accustomed to having certain crops all year round so out of season our favourites can be imported from around the world.

Of course there are many more exotic fruits and vegetables readily available for us to try and thanks to the enormous advances made in transport and refrigerated storage, growers can market their produce on a larger, far reaching global scale enabling supermarkets to sell more and more of the lesser known varieties of cereals, pulses, nuts, fruit and vegetables.

Don't be afraid to try something new from a reputable supplier - you might like it, even if you're a bit conventional towards foods as I am. Just follow the instructions on the packet – some pulses for example if purchased in their dried state do need to be soaked before use.

Have you come across any of these? Couscous, (of the *Poaceae* (cereals) family) made from small balls of steamed and dried wheat, more like pasta than rice.

Vegetable squashes and pumpkins (related to the cucumber *Cucurbitaceae* Family) come in a variety of interesting shapes and sizes. They do however need to be cooked if you can cut through the very hard outer surface. In many countries containers, called gourds, are carved out of the strange shapes of squashes.

There are some rather unusual root tubers that generally produce carbohydrate food. Yacon (of the *Asteraceae* (Daisy) Family) looks like a potato, grows like a Jerusalem artichoke and tastes something like a pear! Use it peeled in salads or to add a crunch to a stir fry. Mashua (of the *Tropaeolaceae* (Nasturtium) Family) are peppery tubers – cook them like potatoes. Then there are *Vitelotte noire*, gourmet French purple potatoes, yams and sweet potatoes.

Cassava (Tapioca, of the *Euphorbiaceae* Family) a nutty flavoured tuber is also a major source of carbohydrate but this is not one for the faint-hearted as if not prepared or processed correctly the cassava plant can produce cyanide, a deadly poison - so do buy from a safe market source such as a supermarket or specialist professional suppliers!

Mizuna (of the *Brassicaceae* (Mustard) Family) is a peppery salad leaf, Mitsuba (of the *Umbelliferae* or *Apiaceae* Family) a Japanese wild parsley used in miso soup, salads and sushi.

Most edible vegetable oils come from crushing the seed of the plant. Common examples are sesame seed oil, sunflower oil and many others such as corn, rape seed and mustard oils. Olive oil, just to be different, is produced by crushing the whole olive fruit.

Olive trees can live for hundreds of years

Edible oils are usually used for making cooking oil and fats in margarine, for toppings such as mayonnaise, salad cream and many dairy products.

Palm oil is used in many products, both edible as in margarine or ice cream for instance, and inedible, such as cosmetics and detergents. Sadly, this is causing much environmental conflict as huge areas of tropical rainforest have been cleared to plant the necessary palms, destroying the habits of many endangered species of animals as well in some cases as evicting human forest dwellers

Unusual oils also come from cocoa, cotton, hemp, flax, mango butter and shea fats but are mostly used as industrial products which will be covered in the next chapter on Plants for Industry.

Originally from the Spice Islands, which today is mainly the South East Asian Indonesian region, the flourishing trade spice provides the whole world with exotic ground spices. Many spices are now grown in tropical areas including India. Variety is the spice of life they say, and there certainly is a large variety of different spices to tempt our taste buds and add flavour to our food.

Spices are made by drying the part of the plant to be used and then grinding it into a fine powder. Nutmeg is the seed of the *Myristica* Family tree and mace is derived from the seed covering (husk) of that same tree

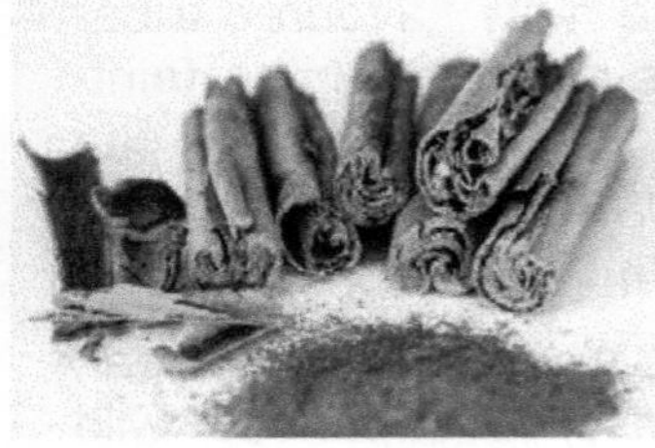

Not just for Christmas!

Cinnamon is obtained from the inner bark of the Cinnamon Family tree. The bark is removed and dried and then rolled to make a tube. It is marketed in the form of short round sticks as well as powder. It is widely reported to have many health benefits. I just know that my wife always adds it to her homemade apple pies and they always taste good.

Ginger and turmeric (the main spice in curries) are the dried rhizomes of herbaceous perennials in the *Zingiber* Family. These plants are considered to have medicinal benefits, ginger aiding digestion in particular.

Saffron comes from the flower of *Crocus sativus* of the *Iridaceae* (Iris) Family. It is produced by grinding down the three dried stigmas of the plant, and is the most expensive spice in the world. Iran is currently the main producer. Saffron was once grown commercially in Saffron Walden in Essex, that's where the town got its name from. Its production in the UK is now being trialled again by an independent farmer, but as it takes 250,000 stigmas to make half a kilo of the spice it may be some time before the project takes off.

So spice can be obtained from a tree, from a root and from part of the reproductive organs of a very small flower. Of course, there are many other spices, all readily obtainable from your local supermarket – it's fun to experiment.

Home-grown herbs are another delicious way of adding flavour. It is good to grow and pick your own, but of course, herbs are also sold chopped and dried for instant use. Mint, rosemary, sage and thyme all belong to the same family of aromatic herbs (*Lamiaceae* the Mint Family) and it is the high level of volatile oils contained within them that gives off those mouth-watering smells. The Mint Family is prolific in producing new leaves – the more you cut, the more they will grow. My particular favourite is basil, although like me, it is rather sensitive to cold weather!

A tasty addition

Edible bulbs, onions and garlic, and edible fungus, mushrooms, all add a wonderful taste and aroma to our cooking. Someone once told my wife that if she was late in preparing the evening meal and hubby was on his way, just throw some onion or garlic into a pan with a good knob of butter set on a low heat, and when he arrived home the resultant smell would forestall any argument about the late meal .

And a colourful plate

Just to make your dinner plates look perfect, some flower petals are edible.

Nasturtium is a good example, although as this particular plant tends to attract blackfly, I wouldn't be too happy eating them straight from the garden without a thorough washing first.

Also check before you eat exactly which part of the plant you can eat as the flowers may be edible but the stems may be poisonous, you have been warned!

Some very refreshing drinks can be made from flowers. Elderflower cordial is a particular favourite down in the West Country in the UK although I have to say the amount of sugar required seems to be rather excessive, sugar being the 'food baddy' of the moment.

I have talked about everyday foods and about exotic foods ….

Now for those absolute essentials: Where would we be without our morning 'cuppa' be it tea or coffee.

Tea is made from the dried leaves of *Camellia sinensis* of the *Theaceae* (Tea) Family, an evergreen shrub from the Asian and Indian subcontinents. It wakes us up, calms us down – there's nothing like a decent cup of tea to soothe the troubles of the world. How many cups a day do you drink?

Black or white?

And then there is coffee from the South America coffee bean plant, Coffea from the *Rubiaceae* Family. Not really a bean, coffee is made by grinding the seed along with its endosperm. Good to drink at any time, particularly after a meal I find, and often used to give a quick stimulus, but don't drink too many late at night, or it won't be so easy to get off to sleep.

And if those 'cuppas' are not sweet enough, sugar cane, a species of tall perennial grass, accounts for 80% of the sugar required by those sweet toothed nations around the world.

Hops are the flowers of the climbing vine plant, *Humulous lupulus* of the *Cannabaceae* Family – yes the same family as cannabis. The female cone-shaped flower is used to give beer its bitter taste.

Fancy a beer?

Mmmm - Chocolate

The cocoa bean is the seed of *Theobromo cacao* of the *Malvaceae* Family found mainly in West Africa. A cup of cocoa, soothing before we go to bed, or that bar of solid brown deliciousness when nothing else will do.

Most of our edible plants are Angiosperms, the flowering plants. As always, there are exceptions and today there is an interest and a market in foods derived from non-flowering conifers and ferns.

ᐅelicious pesto

Certain conifers, particularly pines, produce edible nuts such as the pine nuts from the species *Pinusmonophyla* or *edulis*. A simple pesto sauce made with basil, parmesan, olive oil and ground pine nuts is a favourite to flavour pasta.

Pine needles infused in hot water make an usual tea, but don't forget to strain before drinking!

The berries of the Juniper tree a member of the Cypress Family (*Cupressaceae*) are the softest conifer cones, and are famously used as flavouring for gin, but are also used to give added flavour to casseroled meat dishes, marinades or stuffing. They can also be added to fruit cakes.

Of the many species of fern only a few are edible. The fronds are picked early in the season before they unfurl and are often given the name fiddlesticks as they look like the scroll on a stringed instrument I personally wouldn't recommend foraging – if you wish to try this rather unusual vegetable, seek out a reputable supplier. I understand that they are used in Asian cuisine and also in Northern France.

Seaweed anyone?

Seaweeds are unfortunately named in my view, but we all have our own definition of what a weed is. Botanically seaweed is described as a multi cellular plant which is an algae. It is familiar to us as the kelp, rockweed and gulfweed found growing on seashores, salt marshes or submerged in the sea

There are some useful aquatic food crops grown in other parts of the world. Many herbs can be grown as seeds then grown under water. An example of this is the red lettuce grown off the coast of Italy.

In Africa floating rice crops can be found growing on the River Niger as well as on lakes in some tropical areas. Grown in water this way helps protect the crop from hungry predators, a good organic process.

Of course, plants not only produce our food. They are used in manufacturing, for timber, clothing, natural fibres, the dyeing industry, fragrances, the rubber industry and very important medicines.

What would we do without them?

Chapter 4 Plants for Industry

Could Stone Age man ever have foreseen the amazing variety of everyday products that could be made from humble plants? Doubtful, but they came up with a lot of good ideas - they needed to in order to survive.

Necessity is the mother of invention so it is said and science is the study of the natural world through observation and experiment. How many experiments our ancestors must have undertaken to provide us with what we take for granted today. What wonders they worked with the abundance of plants nature had provided.

Today, of course, experimentation continues. In the industrialised world that we now live in we have been able to move on, investigate further and maybe use synthetic materials to replace some, but by no means all, of the materials originally only available from plants.

Damaging our wild life and environment

There are, of course, some synthetic products that are maybe not quite so good for the planet. Plastics for example - good for the myriad of useful applications we wouldn't be without, bad for the environment when disposed of thoughtlessly and downright ugly when washed up on the shoreline or littering our highways and byways. We urgently need biodegradable plastics to conserve and protect our oceans and wildlife habitats.

Many countries around the globe do still rely heavily on natural products. Natural bio fuels for vehicles are working well in some large countries such as Brazil to avoid expensive oil imports.

With world-wide calls for a reduction in carbon emissions to reduce potential global warming, we have to strike a balance between natural and synthetic, at the same time making sure the natural products are not over-harvested so that there will always be a ready supply for future generations and that the world's wildlife does not suffer because of our greed.

So many uses

Wood - timber - was probably the first plant based building material used by man. Wood a natural product, one which we cannot improve upon, or can we? Wood was used to light fires to keep caveman warm, and is still used by many today as a domestic heat source – an open fire, that cosy log burner or even wood chips to feed a biomass heating system.

Trees come in many shapes, sizes and colours. There are trees that provide hard wood, and those that provide soft. Too many to mention, but each has its own particular use.

Fast growing pine is used for many of our everyday basic wooden products, but when something special is required there is a large variety of different woods to choose from.

Mahogany, hard wood for furniture

Mahogany comes from tropical hardwood trees in the *Meliaceae* Family grown mainly in South America. It is a straight-grained long-lasting wood with a deep rich reddish-brown colour. It is used to make musical instruments, the violin, for example, for furniture and also in a marine environment in the construction and fitting out of boats.

Cedarwood comes from many species of non-flowering cedar conifers but sadly many of these have been over-harvested to obtain the fragrant, insect repellent wood.

Oak timber (of the Beech - *Fagaceae* Family) is a hard wood with many uses in the construction industry. What would that idyllic country cottage be without its old oak beams, although I am sure people used to be shorter!

Oak takes a long time to mature. The first acorns take 20 - 30 years to appear, and a tree will often have been growing for 150 years before it can be used in construction. I wonder how many of these wonderful old trees were used to build Henry VIII's Mary Rose. Just to think that the remains of the wreck which can be seen today at Portsmouth UK were probably obtained from trees growing over 700 years ago.

Oak is used for ageing barrels for wine and spirits. Cork oak (*Quercus suber*) an evergreen tree from the Mediterranean provides the cork stoppers for our wine bottles

From cradle to grave (think cots and coffins) we are surrounded by products created from wood.

The building blocks of houses - from the rafters above to the joists below. Floors, doors, windows and bannisters. From tables and chairs and high-end kitchen units to lowly wooden spoons and old-fashioned clothes pegs – we need wood.

From church pews – are they made so hard we don't fall asleep during the sermon? – to garden fences and railway sleepers - we really need wood.

From loo rolls to tea bags, newspaper to wallpaper, pallets and packaging, wood chips to sawdust. I challenge you take up a wooden pencil and a piece of paper (wood pulp) and see how many wooden products you can write down in 60 seconds – I am sure you will be surprised just how many there are.

We really do need wood but, of course, for it to be grown in a sustainable way.

Natural or synthetic, we need to strike a balance.

Where the very modern chemically made materials of plastics, rubbers and fibres which have appeared since the Second World War are based on man-made polymers, guess who got there first with polymers – man or plants. Naturally, it was the plants.

Cellulose is a plant polymer created from joined up molecules of the organic sugar chemical, glucose, produced as a result of photosynthesis (covered later). Cellulose is the building block of the cell walls of green plants. There are cellulose fibres in wood, in bark and in leaves – really in any living plant material. Lignin, an organic substance found in certain plants, binds cellulose fibres together, strengthening the plants and enabling them to stand erect. A very complex polymer this one, with a chemical equation not for the faint-hearted!

Plant cellulose can be turned into paper, film material, plastics and amazingly, explosives.

Different plants produce different types of cellulose. A good example is the cellulose taken from cotton, which, when mixed chemically with nitrates, makes the explosive guncotton! Cotton cellulose, mixed chemically with camphor makes celluloid. Celluloid is difficult and expensive to produce, as well as being highly flammable, and is not therefore widely used today. One celluloid product that has survived the test of time though is the good old table tennis ball.

Hemp has been around for many thousands of years, and it was one of the first plants to offer up its fibres for spinning into simple clothing. Industrial Hemp is a very fast growing form of *Cannabis sativa* (*Cannabaceae* Family). Fortunately it does not have the effects that we may associate with the drug Cannabis but it does have a myriad of uses in the industrial world. The plant grows tall quickly up to about 5 metres, and contains large amounts of cellulose.

Just love this rope

Hemp has been made into paper, rope and sailing canvas. It is used to insulate buildings and in acoustic panels as sound-proofing. It can be refined to produce oil, wax, resin and plastics. It is used in animal bedding and the seeds are a real treat for birds

It earns its place in the bathroom, being used in towels, in moisturisers and as a composite material in the manufacture of sinks. Plumbers even use it to stop the odd leak. All in all a remarkable plant, one which does not deserve the bad publicity of its notorious cousin, *Cannabis indica*. In some parts of the world it is illegal to grow certain species of Cannabis without a Government licence – it is in the UK.

Bamboo (*Poaceae* Family) is a very large and mostly hardy grass which given the right conditions can grow up to 3 feet in a day! Well I am sure ve all use bamboo canes for stakes in our gardens but there are many ther, perhaps surprising, uses of bamboo. It is used in Asia as caffolding and in the construction of simple buildings, especially where looding is an issue and dwellings need to be raised above flood level.

So many uses for bamboo

Bamboo is used to make cane urniture, wooden chopping boards nd utensils, and even in the nanufacture of clothing.

Vho would have thought of socks nd tee shirts from bamboo?

Bamboo fibres obtained by a process of steam extraction have been onded to cotton yarns and other natural fibres such as the hemp plant. These natural fibres are being used more and more to produce eco riendly bio-degradable fabrics.

These valued by-products may contribute significantly to the major efforts o control climate change because many of them will be carbon neutral as hey are naturally grown products which absorb carbon dioxide. We need lot more of these carbon dioxide guzzling plants to help reduce the ffects of global warming.

Natural pigments, dyes, cork, waxes and resins all come from plants. Resin is a sticky non water soluble substance that is secreted by trees. It s extracted as a liquid but sets hard when exposed to air. Frankincense nd Myrrh famous from biblical times are resins extracted from trees in the *Burseraceae* Family. The oils of Frankincense and Myrrh are feted for their calming, stress reducing properties and as an aid in the reduction of pain nd inflammation.

Research is currently being undertaken as to whether Frankincense may help eliminate the spread of cancer cells and whether as an anti-inflammatory it may help with the problems associated with arthritis. It is good that there is constant research taking place in the field of medicine, but I must emphasise my belief that professional medical help should always be the first port of call.

Linseed oil, obtained from the seed of Flax *(Linum usitatissimum)* was historically used as a preservative for wood, famously for cricket bats, in concrete and in paints and varnishes. It was also used in the production of linoleum a rather old fashioned floor covering product Linum (*Linaceae*). It does have drawbacks though, being somewhat sticky, slow to dry and quick to attract mould. Technology moves on and today there are better products to be found on the market.

The fabric, linen, is made from the fibres of the flax plant. The linen industry is massive worldwide, fashioning a wide range of clothing, bedding, curtains and the like. I have to say my wife is not very keen – she does not go for the 'creased' look of many linen products. What is interesting though is that the by-products from the linen industry are processed into a pulp used for bank notes. Funnily enough, my wife says she doesn't mind if she finds her money 'in creases'!

Some research has also shown that flax seeds contain a type of omega-3 fats which can be beneficial in reducing high cholesterol levels. Not sure that crushed seeds will necessarily be the first choice for the ever increasing number of people taking statins in tablet form, which would seem a far easier option

Cotton bolls

Another fabric with a plant base is Cotton Gossypium of the (*Malvaceae* /Mallow Family) similar to Hibiscus. The cotton plant is grown in tropical and sub-tropical areas of Japan, Australia and USA. A green shrub about 3 metres in height, it is mainly cellulose, grown for the cotton bolls 'fruits', harvested and pressed into bales, then spun, dyed and woven or knitted into fabric for furnishing and clothing.

Cotton seed oil is used in the production of margarine and some salad dressings. Amazingly cotton seed oil exceeds the production of soybean oil from the *Leguminaceae/Fabaceae* Family, canola oil (a cultivar of rapeseed from the *Brassicaceae* Family) and corn oil (or maize from the *Poaceae* – Grass Family).

The rubber tree (*Hevea brasiliensis*) of the *Euphorbiaceae* (Spurge) Family is grown in tropical climates, originally in Brazil but is now grown particularly in Malaysia, India, Ceylon and Thailand.

Plants in the *Euphorbiaceae* Family produce a highly irritant sap so caution must be taken when cutting back any of the fast growing Euphorbia species that you may have in your garden. The sap from rubber trees in this family contains a specific organic chemical, isoprene. Together the units of isoprene form another chain of polymers which become the building blocks of natural rubber.

Tapping for rubber

Further research is taking place with different species such as the fig rubber tree *Ficus elastica* and the *Palaquium gutta* which produces sap known as guttapercha. This latex sap takes a more thermoplastic rigid form when exposed to air but it can be heat treated and shaped, and has been used successfully in dentistry particularly for root canal work. Ouch!

Naturally grown elastomer (rubber) from the gum rubber plant has different properties to synthetic rubber that is produced chemically. Natural rubber is used for waterproofing and insulation, whereas synthetic rubber is not an insulator but is preferred for vehicle tyres as results show that the heat build-up with a synthetic product is significantly less when compared to natural rubber.

As a polymer science graduate I found rubber and plastics technology really fascinating. I remember sheets of raw rubber being fed into cylindrical crushing mills, compounded with sulphur for vulcanising (turning the viscous liquid to a solid form) and the combination being put into moulds and heat treated to obtain the required shape.

From knicker elastic to condoms, wellie boots and rain macs, rubber balls and trampolines – we've all had our ups and downs with rubber.

Would you believe natural rubber latex is now being used in making artificial grass? As a happy gardener who spends many pleasurable hours mowing my lawn, I simply cannot understand why anyone would want to use this product! Even cruise liners are growing natural grass for picnicking on the upper decks – although I must say I thought this a little weird.

Cosmetics and what my granny used to call 'smellies' are not really my thing, but I remember that Aunt Flo always liked a bar of lavender soap or talc at Christmas time, and I have seen similar products smelling of roses or lily of the valley. These flowers are highly fragranced to attract pollinators – not sure who Aunt Flo was trying to attract with her not too subtle smell of lavender.

I've just planted some lavender – the bees love it

Lavender oil, it is said, helps to reduce stress, improve sleep, slow aging and improve skin complexion and reduce acne – perhaps Aunt Flo was on to a good thing.

Fragrances, scents, perfumes - it takes someone with a good specialist and commercially experienced nose to choose the perfect mix of ingredients to make that enticing smell which we hope will attract the opposite sex.

All I can say is that nature provided us with the basic palette from which to choose. Do you favour a light floral scent, something with a citric base or a more intense and oriental spicy note?

Oil from avocados and *Aloe vera* is used in skin moisturisers, the elder tree is used in the production of creams and lotions, particularly against the effects of too much sun, and very simply two slices of cucumber strategically placed refreshes tired eyes and relieves puffiness.

Here comes the bride

The crushed leaves of the tropical Henna plant from the *Lythraceae* Family have been used by women for centuries to adorn their bodies. As a dye for hair, to colour fingernails, but probably the most well-known application is when it is formed into a paste and used to paint intricate non-invasive tattoos known as Mehndi on the hands and feet of brides to be in many Middle Eastern cultures.

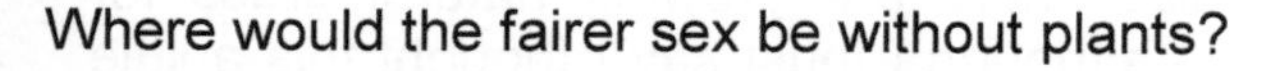

Where would the fairer sex be without plants?

Finally, talking about the sweet things in life, sugar cane (*Saccarum officinarum*) is another grass which contains around 17% sucrose and is grown in the tropical zones of South Asia and Brazil. It produces approximately 80% of the world's sugar.

As well as the sugar, currently being blamed for high levels of obesity, fibres from the actual cane can be used to make mats, screens and many similar products.

As sugar cane cannot be grown in areas with a colder northern climate, a root crop product, sugar beet, which can be processed into sugar, is grown instead, particularly in the East Anglian region of the UK.

There have been recent problems with sugar beet seed germination, possibly not helped by warmer spells in winter and rotting of the seed and roots. Could this be nature's way of telling us we need to cut down on sugar consumption?

My two passions in life are gardening and science – how well they go hand in hand. We have learned so much from plants to help us in our scientific quest for a better world. Let's hope that our future experiments can be achieved in an environmentally friendly manner and so retain the balance in our ecosystems.

Chapter 5 Kill or Cure

'It will either kill or cure' – a strange expression I have always thought, but rather apt when it comes to the first attempts made by man to use plants for medicinal purposes.

Latest news - 'Hunter Gatherer killed by wife'

Imagine the scenario – caveman returns home after a hard day's hunting. 'Take this dear' says his wife, offering him one of her potions to ease his aching bones, cuts and bruises, but oh dear, she hasn't picked the right plant, or used rather too much in the concoction, and that is the end of poor Freddie Flint-Rocks.

Records show that plants have been used for centuries to help with day-to-day ailments.

Particularly as an aid to digestion; to 'help you go', Flaxseed was used for constipation, to 'stop you going', Bilberry was used for diarrhoea, Feverfew was used to alleviate headaches, Cloves used for toothache, *Aloe vera* for burns. BUT - as we now know, many plants are poisonous, and I suspect this may well be where the term 'Kill or Cure' came from. Without the aid of a poor mouse to test these early remedies I am sure there were quite a few disasters along the way.

Certain plants have developed the defence of poison simply to deter predators from devouring them. Others can be affected by viral and fungal attack and by the absorption of toxic compounds from contaminated soil.

It is best to check out what poisonous plants should be avoided in your own particular surroundings, but there are some every-day surprises we should all be aware of. Leaves from a potato plant, along with any actual potatoes that are beginning to turn green, apple pips, cherry leaves and stones and lupin seeds could all create a digestive upheaval. Certain citric fruits can prove to be upsetting for the stomachs of our family pets, so best keep these tasty treats for ourselves.

In mediaeval times superstitions and prejudices were the norm and blame was placed on anyone, anything and everything for causing illness when

what was really needed was a better standard of hygiene and some clean water. Running water was present in reputable treatment centres such as Monasteries and Cathedrals although generally wine or beer was used as antiseptic and to prevent the spread of germs. Perhaps the lucky ones got to drink the left-overs?

The more educated members of society at that time tried with varying degrees of success to develop 'medicines'. Some remedies were even tried against the dreaded bubonic plague which spread rapidly through the bite of an infected flea, but there are no reliable records of any success other than the natural immunity of those few lucky survivors.

Nonetheless, a wide range of herbal treatments has been recorded over the years and indeed 'hospitals' or 'infirmaries' were long ago established to delve further into how to eliminate various ailments and diseases, whilst at the same time relieving the poor patient of much if not all of the cash in his pocket. The modern day NHS may have its problems but we would be a lot poorer without it.

The evidence base of these herbal medicines would have been evaluated very simply on whether the recipient, the patient, responded with an improvement in health in a reasonable time with some hope of remission of the condition. It remained to be seen whether the 'medicine' prescribed was indeed a cure or whether the luckless patient failed to make progress, or sadly died.

Many herbal medicines used over time are now accepted as having a degree of known success as well as being the forerunners of the modern day pharmaceutical medicines produced industrially and prescribed by the medical profession.

As a scientist I will always be open minded on the choice of herbal against manufactured medicines if patients indeed are fortunate enough to have the choice. Sadly in poorer countries today where many conventional drugs are unavailable people may have to rely on herbal medicine but do so knowing their ancestors may well have used them historically with some degree of confidence.

Herbal remedies have been sought and used over hundreds of years for anxiety, for relaxation and as an aid to digestion. Wound and skin infections following some of the terrible injuries received in ancient battles may have relied on herbal treatments to provide some temporary pain relief but it is beyond the scope of this book to cover herbal medicine in depth.

In the UK in London there is an admirable hospital, previously known as the Royal Homeopathic Hospital and now renamed the Royal London Hospital for Integrated Medicine. The Royal London continues to offer its excellent world-wide reputation for Herbal and Complimentary Medicine alongside conventional mainstream medicines. It is good to look at all options.

Please do not be tempted into trying to use plants as a medicine without seeking professional advice. Many plants have poisonous properties and can inflict more damage than we realise. The leaves of rhubarb for example are toxic with oxalic acid which caused a few problems initially in the First World War with people thinking the whole plant was edible. If you wish to try a herbal remedy, buy the appropriately manufactured product from a recognised source and look for the THR mark (Traditional Herbal Registration). Under no circumstances should you try picking medicinal plants yourself. The consequences could be very unpleasant. If you are really keen to 'pick your own' there are now trained foraging experts who are available to lead a party of foragers.

Happily there are a few tried and tested natural remedies that are safe for emergency use.

Naturally useful

Aloe vera from the *Asphodelaceae* Family is a good plant to keep on a kitchen windowsill. The sap it contains soothes minor burns and is a natural antiseptic. Good also for the after effects of sunburn and bruising and as a relief for those insect bites you may well pick up while gardening. It can also act as a skin moisturiser.

Keep young and beautiful!

Oenothera biennis, Onagraceae family (commonly known as Evening Primrose) plants have been known for centuries for the high quality oils they produce which contain a valuable fatty acid not found in many plants and not manufactured by the human body. They are said to have many benefits as an herbal remedy and are particularly used to alleviate skin complaints.

A native North American plant, widely naturalised in both temperate and tropical regions, in the UK Evening Primrose oil comes mainly from organically grown plants from sources which have been certified by the Soil Association. They thrive happily in most soil conditions and like full sun but as the name would suggest only open their flowers as the light begins to fade.

Tasty and good for you

A couple of other natural remedies to keep in the cupboard are ginger and garlic. A small piece of stem ginger seeped in hot water is a useful digestive aid.

Garlic, albeit making the breath somewhat offensive, has many medicinal benefits, including counteracting, it is claimed, colds, high blood pressure and cholesterol. Personally I would not want to eat a large amount of raw garlic, but when my wife cooks with it, I can always be found sniffing appreciatively in the kitchen.

Herbal teas have been produced and marketed as a natural way to combat insomnia, help digestion, bloating and constipation. Try a cup of chamomile tea before bed to help sleep, probably better than a late night coffee, and ginger or peppermint tea to aid digestion. Tea made from the *Echinacea* plant of the *Asteraceae* Family is said to help ward off colds. It may be worth trying some of these natural remedies from a reputable source if you believe they could help you. Genuine tea comes only from the *Camellia sinensis* of the *Theaceae* Family. Similar tea products are usually known as 'infusions' and cannot correctly be called 'tea'.

A herbal preparation, known as Mexican Yam, is manufactured using chemicals obtained from the Wild Yam (*Dioscorea villosa*) of the *Dioscoraceae* Family. This herbal remedy is said to help with cramping period pain, relieve morning sickness in pregnancy and offset some of the symptoms of the menopause – one for the ladies then! I have also read it can be used for nervous excitement and flatulence!!

When I was very young I well remember an elderly aunt telling me that both fire and water were very good servants but very bad masters. I puzzled this for a time until she explained that when humans were responsibly in control these two elements are very useful and necessary but when fire and water take over things can get very bad.

I guess this also applies to some plant species so here are some typical examples of plants that thrive in our gardens, are of great interest medicinally, but can also be fatally poisonous .

Beware extremely poisonous!!

Deadly Nightshade (*Atropa belladonna*) from the *Solanaceae* Family, for example, is as its name would suggest, very toxic, all parts being poisonous.

It actually comes from the *Solanum* Family, as do potatoes and tomatoes, and although we eat these, the stems and leaves of both of these plants are toxic to both humans and animals.

Atropine used to dilate the pupils before an eye examination comes from the Belladonna plant.

Please don't - Tobacco is bad.

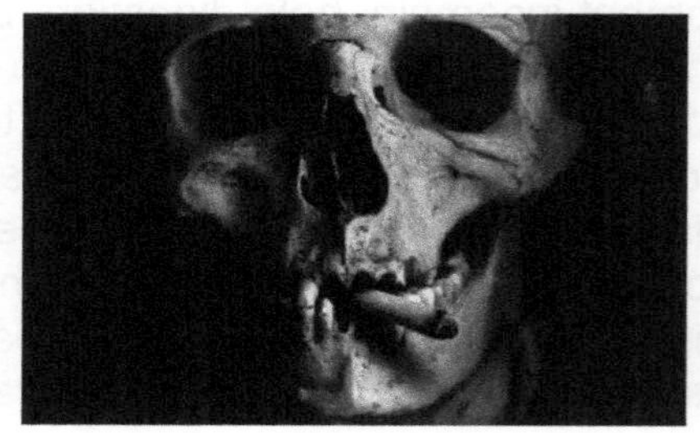

Also from the *Solanaceae* Family comes tobacco, made by curing the leaves of the plant. Tobacco, which is then mainly used in the production of cigarettes. Tobacco, which in 2008 the World Health Organisation blamed for the single greatest cause of preventable death.

The smoking of cigarettes is highly addictive and it is very difficult to quit; I can only say to any youngsters reading this, please do not be tempted to try it, it will only shorten your life.

The Foxglove, with its spikes of bell-like flowers was officially named by Carl Linnaeus as *Digitalis purpurea* (the Latin translation of purple finger) – a good description of the plant. A member of the *Scrophulariaceae* Family, the tall and often invasive Foxglove is attractive to bees and other insects, but all parts of the plant are highly toxic for human consumption. Foxgloves are, however, the source of digitoxin, which in drug form provides a heart stimulant which has been used by man for over 200years.

Poppy seed heads

The Poppy (Papaver somniferum of the Papaveraceae Family) the symbol of remembrance for all who have lost their lives in war, has been used since ancient times as a pain killer and also a recreational drug.

Poppies are grown commercially in the UK to provide pain relief in the form of codeine and morphine. I expect many of us may have taken a tablet or two of codeine to help with head or tooth ache; morphine is a much stronger drug, administered only by medical professionals when severe pain relief is required.

Heroin, also obtained from the poppy is, of course, highly addictive and an illegal drug. Used sensibly the poppy can provide the means of good and effective pain control, but sadly for many the misuse of the poppy can cause an unwelcome and untimely death.

It is the 'goo' from inside the seed pods of the Poppy that provides the substance for the basis of these narcotic drugs. Sometimes known as 'black or brown tar' the dried latex or 'goo', an illicit opiate, is a crude form of heroin, and highly dangerous. A higher quality, but nonetheless illegal opium is obtained by processing the substance.

The actual seeds of the Poppy, rich in minerals and dietary fibre, can be added as flavouring to cakes, pasta or even ice cream. They are said to help sleep, relieve constipation and lower cholesterol, but if as an athlete you eat too many, you may find that you fail a drugs test!

Discovered from ancient texts (not the sort received via your mobile), it has been learned that willow bark of the *Salixaceae* Family was used more than 1500 years BC as a pain relief, reducing inflammation and swelling. A thousand years later Hippocrates was still recommending willow for pain relief in childbirth. I wonder really how this came about. Were women told to bite on a piece of bark when pain became unbearable and found that miraculously it helped?

Nearly two thousand years later aspirin was produced and marketed. Now the wonder ingredient salicylic acid discovered in willow bark is synthetically made and added to corn starch and water to make the aspirin tablets we know today.

Make mine a double!

Another drug originating from the bark of the Cinchona of the *Rubiaceae* Family is Quinine used to combat malaria. This is one where medical supervision is definitely needed as it can cause very unwelcome side effects, although a little is used to flavour tonic that goes well with a glass of gin, ice and a slice of lemon or lime as the British Indian Army discovered!

I am sure most of you will know of the Periwinkle or Vinca plant (a member of the *Apocynaceae* Family). This fast growing plant with its pretty purple flowers can be found in many gardens – in fact if left unchecked it can become something of a nuisance - but there is one particular species *Catharanthus roseus* (commonly known as Madagascar periwinkle) which has been found to have life enhancing properties in the treatment of cancers.

The Madagascar Vinca has long been used as an aid to reducing sugar levels for diabetics, but in the second half of last century a remarkable discovery was made. The alkaloids (chemicals produced by a plant which are highly toxic) extracted from the plant were found to have reacted against cancerous cells which were reproducing in mice.

Vinca plants grown in tropical and subtropical areas are today used with significant results in chemotherapy given to patients with certain forms of cancer. Following further research it has been found that the alkaloids form a derived compound which has been given the name Vincristine which remarkably has been used with great success in the treatment of childhood leukaemia. There are reports that suggest the survival rate may have risen from 10% to 90% which is immensely encouraging and a very good reason to be grateful both to plants and the scientific research being undertaken.

The English Yew Tree (*Taxus baccata* of the *Taxaceae* Family) – you know the one, dark and rather forbidding, often to be found in churchyards – also contains some really useful chemicals which have proved to disrupt the process of cell division in certain forms of cancer. Most of the chemicals are to be found in the needles of the Yew and as these plants need an annual clipping, a constant supply is readily available.

Of course, cancer research is on-going, but the fact that the toxic chemicals derived from plants used in chemotherapy can potentially provide a cure for a condition when once there was little hope of survival I find a truly amazing scientific discovery.

There are probably around 120 different chemicals that can be extracted from plants. Some products are still made by harvesting the relevant parts of the plant, whether it be the leaves, flowers, roots, berries or seeds, to produce naturally the chemicals needed to make the particular drug required.

Today, however it is easier and cheaper to synthetically engineer some, although not all, plant chemicals, but if it had not been for the perseverance and dedication of the first botanists into the science behind the plants they were growing, life would be very different for us

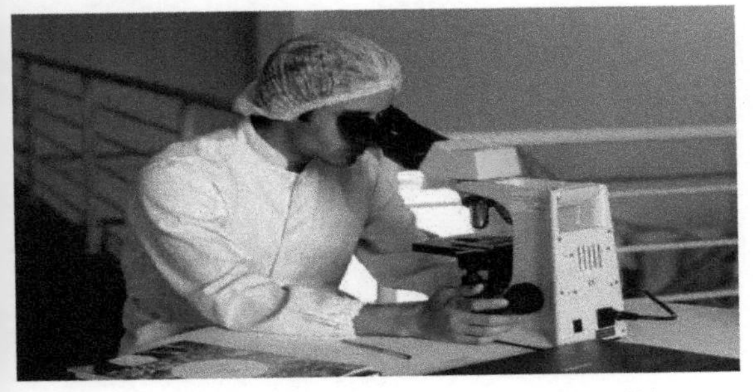

Science and gardening go hand in hand

Some natural herbal products are regulated by law in certain countries, for example, Germany. In others, such as the USA and others parts of the world, they may be freely available.

Lists of standard drugs and chemicals, along with their clinical use and plant source, which I find make very interesting reading can be easily accessed on-line.

Botanical Latin is not for everyone!

Chapter 6 Name that plant

Many plants are known by their common names – perhaps think of these as 'nick-names' - but they also all have 'proper' Latin names, which are known as their botanical or scientific names written in ***Italics***

This is confusing to many gardeners so I'll attempt to set the record straight and clarify the situation.

Common names will normally be given in the language of the country where the plant grows, so are of little use world-wide. In many countries, such as the UK, different regions may have different common names for the same plant. Some very strange and amusing names for the plant botanically named *Sempervivum* are: the 'Houseleek', 'Live Forever', 'St. Patrick's Cabbage' 'Hen and Chicks', and very strangely 'Welcome home husband however drunk you may be'! I am not sure my wife would be so forgiving! As you can see common names can be very descriptive locally and are especially popular for wild plants.

Some plants have been given rather unpleasant nick-names but for a good reason. *Hellebore foetidus*, known also as Dungwort, but often called 'Stinking Hellebore' is named for the smell of death of its crushed leaves. I would advise against treading on this particular plant whilst gardening.

Birds love Sorbus berries

Even worse the 'Bastard Service Tree' named after the popular tree the *Sorbus aucuparia*. I wonder how it got such an unfortunate name !.

Sansevieria, a difficult-to-kill houseplant, has several nick-names, but my favourite has to be 'Mother-'in-law's Tongue'. Sadly, this plant is poisonous to domestic pets.

An unusual shaped flower

Arum maculatem commonly found in woodland, has many and varied nick-names, one of which is the Lords and Ladies Plant, because of a perceived likeness of parts of the plant to parts of the male and female anatomy which should perhaps remain out of sight.

Domestic food, fruit and vegetables are unlikely to be marketed with botanical names. What housewife (or husband) shopping in a greengrocers is going to ask for *Lactuca sativa* (a lettuce) or potatoes *(Solanum tuberosum*).

Botanically or scientifically the way we name plants today is based on what is known as the Binomial (two name - genus and species) system created around 300 years ago by Carl Linnaeus, which has generally stood the test of time. The genus always starts with a capital letter. Think of genus as the surname, for example Geranium.

The species represents a sub-group within the genus e.g. *pratense.* The species always starts with a lower case letter. Think of the species as the forename. So putting together the genus and the species the plant takes on the name *Geranium pratense.*

This system is a more precise way of naming a plant as the species name used can also be descriptive of a plant's colour or shape as well as describing the likely habitat.

I'll give you just a few examples.

Species named nudiflorum describe plants where the flowers appear before the leaves. For example *Forsythia nudiflorum*, where bright yellow flowers heralding spring appear before the leaves of this vigorously scrambling, deciduous shrub.

Species given the name japonica or japonicum are used to describe plants that originated in Japan; surprise, surprise! *Acer japonicum* is a genus of small growing deciduous trees the leaves of which provide a gloriously colourful autumn display.

Rubens is used to describe plants with red colouring. Picea rubens is a red spruce native to eastern North America. It is an evergreen tree, sometimes also commonly known as the yellow spruce. An evergreen tree, which can be known as a red spruce or a yellow spruce – confusing!!

Another mystery is why the *Hellebore niger* (niger being Latin for black) actually bears white flowers!

The Binomial system was readily accepted by scientists not only because common names proved to be hugely variable but also because the very lengthy Greek and Latin names that were originally set down to describe a plant were just too long and cumbersome to use or remember. The Latin language was a popular choice for this at the time with most educational institutions speaking Latin. Nowadays although very few of us even study Latin, this ancient language continues to be the one used for botanical names and is accepted by the International Code of Botanical Literature that sets the rules for Plant Naming.

So exactly who was Carl Linnaeus who created the Binomial naming system?

Carl Linnaeus was born to Swedish parents at the very beginning of the 18th century. His father, who was a clergyman and an amateur botanist, was married to the daughter of a clergyman. Guess what they wanted their son to be when he grew up? It was not to be - Carl had no interest in following his father into the Church.

He did however share his father's love of plants and was fascinated by the many different flowers that grew in the rectory garden where he lived as a small boy. He spent many hours happily following his father around in the garden, constantly asking the names of the flowers; he was a show-off at heart, wanting to be able to impress and to be able to name as many of the plants as possible. His father taught him botany and Latin from a very young age and Carl was able to grow his own plants in the large family garden, before venturing further afield to the nearby meadows and marshland seeking out different species to study.

When he was just a young lad he was sent away to a school for would-be clergymen, but did not excel. His mind was too full of plants and his teachers did not consider him able enough to further his clerical studies at university. All doubted his abilities, except one, a medical doctor, who recognised his love of plants and suggested that Carl follow a career in medicine. Studying botany at this time in history was of great importance if you were to follow a medical profession as all drugs were derived from medicinal plants. Off he went to take his place at university.

Carl was intrigued by the different ways that plants reproduced, writing down his many observations, and for a time at the very young age of 23 he lectured on botany at university. His lectures were very popular and his mother was now content that her eldest son was a success, but I wonder if she ever realised just how famous he would become.

Maybe as you have been reading through this book you have skipped the Latin names I have tried to introduce – not the end of the world. Botanists at work and research would usually use the Latin family name but I have to say the English name is the one generally used by most gardeners. What I would try to impress upon you is the Importance that should be placed on what family a particular plant belongs to.

All plants, from the tiniest weed to the most gigantic of trees, belong to a family. Whether they are flowers, shrubs, trees, herbs, fruit or vegetables they all belong to a family.

It was Linnaeus' suggestion that to simplify matters plants should be placed into 'families' based on the structure of their reproductive organs!! He counted the number of anthers and stigma of each plant he came across and if the numbers were the same he put them into the same family, disregarding variations in shape, size and colour.

Plant Taxonomy (modern day botanical naming) deals with the grouping and classifying of plants into the most appropriate family where similarities in characteristics are just as important as the number of a plant's sexual organs. The introduction of modern day genetic testing now plays a much greater role and consequently name changes need to be made to the old groupings.

Occasionally plants are divorced from one family and placed in another causing a bit of an upset, mostly to gardeners like myself who for years have thought of a plant belonging to one particular family when suddenly it is uprooted and transplanted within another! There are good reasons why botanists make these changes but as with all things scientific it takes time for the lay person to accept and adapt.

At Horticulture College we were constantly being updated on botanical name changes, causing tutors and students alike to raise their eyebrows and scribble furiously in a notebook, promising ourselves to commit the new information to memory at the earliest possible opportunity.

The botanical family name given in Latin always ends in *ACEAE* for example *Asteraceae* is the Latin name for the Daisy family.

There are hundreds of families of Angiosperms (you remember those flowering plants I talked about), containing over 350,000 species. These are all set out in a register, which is officially known as 'The Plant List'.

Plant genera in families can be very similar in appearance - look at the similar shape of the individual flowers on Lupins or *Ulex* (gorse) for example - but they can also be very dissimilar. *Berberis* is a tough, prickly, large growing shrub happy in sun and *Epimedium* is a gentle ground-covering flower which prefers shade but they are both in the *Berberidaceae* Family.

An important fact to remember is that plants in the same species can reproduce with each other. In some cases plants in the same genera can also reproduce with each other, but that is much less usual. Certain pests and diseases are attracted to and may have an adverse effect on plants in the same family. Often plants in the same family enjoy a similar growing environment, having a preference for a particular soil or pH or favouring particularly wet or dry, hot or cooler growing conditions

A botanical family can have just one member, i.e. the *Ginkgo* or the *Araucaria*. On the other hand a family can contain a large variety of differing genera and species plants, which include a mixture of flowers and vegetables or vegetables and herbs, flowers and fruit or flowering trees.

Edibles can be found in the same family as poisonous plants or trees! It is all rather confusing, and a great deal to remember.

To help you I have set out several families for you to familiarise yourself with, including some with a mixture of ornamental, edible and poisonous plants. I recommend that you do some research into the families of plants that you particularly like.

I suppose the first and most important family to mention is *Poaceae*. This is the grass family producing the grain for cereals we need to feed the world.

Grass grows naturally in most parts of the world. Keeping an ornamental lawn in good condition is a challenge for the keen gardener. In summer it either grows too quickly and mowing the lawn becomes another chore, or the weather is too hot and it dries out leaving ugly brown patches. It ticks over well in winter but can become slippery when conditions are icy or muddy and squelchy when too much rain has been an issue. That's the green, green grass of home which shares family membership with the grass crops grown to feed our livestock and those crops cultivated for their starchy cereal grains so important for food world-wide.

The Rose (*Rosaceae*) Family is one of the largest, including the *Prunus* genus of very ornamental fruit-bearing trees such as apricot, plum, cherry, peach and almond as well as the *Malus* genus containing apple and crab apple trees.

Rubus, a genus of soft fruit berries such as blackberries, dewberries and raspberries also come under the *Rosaceae* heading.

Ornamental shrubs in the *Rosaceae* Family include *Cotoneaster, Chaenomales*, *Pyracantha, Kerria, Spiraea, Crataegus* (Hawthorn), *Potentilla, Geum, Sorbaria,* and *Exochorda.*

Don't worry, I'm not going forget the most diverse, the most colourful and the most fragrant member of this family, the rose. Bush roses, shrub roses, climbing and scrambling, every gardener should find space for at least one.

The Pea Family (long called *Leguminaceae* but now re-named *Fabaceae* – I'm not sure why, but I'll try to commit it to memory) contains trees, shrubs, perennials and annuals, as well, of course, as the humble edible pea.

The roots of Pea family members contain nitrogen absorbing nodules and the flowers usually have the typical 'pea family' shaped flowers, characterised by the delightful sweet pea.

Ornamental trees in this family, *Laburnum, Mimosa, Acacia, Genista, Caragana and Gleditsia,* all have the recognisable 'pea family' shaped yellow flowers, although the *Cercis,* which is red, and *Robinnia*, which is white, do not conform to the usual colour pattern.

Shrubs, *Wisteria, Coronilla and Ulex* also have the familiar 'pea family' flower, but annual sweet peas can grow taller, being climbers and usually have multiple colours as well as a strong and distinctive fragrance.

Peas are perhaps the most popular of all vegetables, easily and quickly frozen for long storage or tinned for convenience, but the seeds of the *Laburnum* in this family are very poisonous so ensure children and animals are kept at a safe distance! Best collect and destroy them when shed.

The Buttercup Family – *Ranunculaceae*, thought to be one of the oldest families having evolved from primitive plants, consists mostly of herbaceous perennials, *Hellebores, Aquilegia, Caltha, Delphinium*, *Actaea, Aconitum, Consolida* and *Anemone* to name a few. The family also includes some annuals such as *Nigella* and the important climbers in the *Clematis* genera.

Although attractive to look at, many of these plants are poisonous, and in ancient times *Aconitum* leaves were made into a tea and used in administering the death penalty. Keep them away from young children and pets, even adults can be endangered.

The Buttercup weed, *Ranunculus repens*, can be found growing on open lawns and meadows. I well remember my sister picking one of these shiny yellow flowers and holding it under my chin, telling me in her superior sister voice that if the yellow was reflected onto my chin I liked butter!.

The Mustard Family – *Brassicaceae*. Besides the vegetables, cabbage, turnip, mustard, watercress and horseradish, there are many flowers in this family grown for decorative purposes, such as Stocks, Candytuft, Sweet Alissum, *Aubretia* and *Alyssum.* On the theme of vegetables the Carrot Family – *Umbelliferae* (now re-named *Apiaceae*, possibly because the flowers are attractive to bees!) is a family of mostly herbaceous plants together with the important edibles, carrot, parsnip, parsley, fennel, chervil, dill, celery and angelica. They can produce some impressive tap roots, especially when grown in deep soil.

One to avoid

A point to note is that the very unpleasant giant hogweed is a member of the Carrot family and may cause severe rashes and dermatitis to the skin.

Leave flavouring to plants

Mint family - *Lamiaceae*

As you would expect, this family includes all the varied and widely used culinary herbs, popular for their aromatic fragrant leaves, some for seed and some for edible tubers (including the Chinese artichoke).

It also includes flowers, *Salvia, Stachys, Plectranthus, Teucrium, Nepeta* and the very fragrant shrub, *Clerodendrum.* These also have aromatic leaves and in fact *Nepeta* (commonly known as cat mint) has an almost narcotic effect on cats and if you awake one morning to find your pretty blue plants flattened, you can be sure that you have been paid a visit by one very 'spaced-out' and happy moggie.

Heath Family – *Ericaceae* - All plants within this family require acidic soil in the range 4.5 to 6pH. Try growing in containers filled with ericaceous soil if your garden soil falls within the alkaline range 7-9pH. Ericaceous plants are mostly evergreen shrubs with one attractive and distinctive tree, the *Arbutus* (Strawberry Tree) that bears its flowers and fruits at the same time, late in the year, October up to December.

Ornamental plants include *Erica* (Heathers) and the familiar *Rhododendrons and Azaleas. Enkianthus, Pieris, Gaultheria, Pernettya and Kalmia also fall into this group.*

The family includes *Vaccinium* (edible fruits) which is an important group of berry plants, cranberries, cowberries, bilberries, blueberries and an American fruit, huckleberry.

Lady's Bedstraw – doesn't look very comfortable!

Wild flowers do form part of normal families but are just rather – wild!

They have often been given rather unusual and comical common names. I find it a good idea to take a pocket guide with me when walking in the countryside it helps me to recognise some of these tearaway plants and I have a chuckle at some of the names and how they came about. Birdsfoot Trefoil, Lady's Bedstraw, Frogbit and Birthwort – see how many you can recognise when taking a stroll down a country lane.

The wild flowers in pocket books are set out in separate colour sections which makes identification and naming so much easier. Often wild flowers will have many of the same characteristics as their more cultivated cousins and I am sure you will recognise many that also appear perhaps more sedately in our gardens. Blue Gentian and Foxgloves, yellow Lily and Iris, red and purple Orchids and Poppies spreading a blanket of brilliant crimson.

The fun for me is learning some of the names of the more unusual wild flowers that can be found gracing our countryside. Large Venus's Looking Glass and Sheep's Bit, both members of the Bellflower Family.

Hedge Mustard and Treacle Mustard adding a little heat to the *Brassicaceae* (cabbage) Family.

Then there is Smooth Hawk's-beard, Rough Hawk's-beard, Nipplewort and the rather more-well known and recognisable Dandelion, all of which belong to the Daisy Family (*Asteraceae*).

Studying and trying to track down different species of wild flowers can become a most engaging pastime; see if you can spot some hiding in urban as well as rural locations. Don't forget your pocket book, but please do not be tempted to pick wild flowers, it's illegal in many countries, leave them where they are happily growing for others to enjoy.

While you are out and about looking for that elusive wild flower, try to dentify some of our native trees. There are around 33 species of trees native to Britain which began to emerge around 7,000 years ago as the glaciers formed by the last Ice Age finally melted and the land mass we know as Britain began to move away from its continental neighbours.

Thinking about the following characteristics should help you.

What is the size of the tree? Note its height and the width of both trunk and canopy. Is the tree growing in dry, damp or very wet conditions? Is it evergreen or deciduous?

Look at the shape of its leaves, does it have a rough or smooth bark, or is the bark a particular colour.

Trees in the *Fagaceae* (Beech) Family can be massive with thick trunks and low canopies. They include the Beech (*Fagus sylvatica*) Oak (*Quercus* varieties) and the Sweet Chestnut (*Castanea sativa*) all of which produce nuts.

Catkins are produced by trees in the *Betulaceae* (Birch) Family, Birch, Alder, Hazel and Hornbeam as well as the *Salixaceae* Family, Willow and Poplar.

Evergreen conifers bear male and female cones, and trunks range in size from the smallest dwarf to the gigantic *Sequoia sempervirens* found in California reaching over 100 metres in height.

Until recently there lived a Sequoia that had grown so huge a small car could be driven through an opening in its trunk.

Nature as ever had the upper hand, and sadly following a raging storm, the weakened tree succumbed and is no more. What a great shame – why would anyone want to drive a car through such a majestic tree?

Bizarre

Trees are so important for our environment. They are not just beautiful enhancing our gardens and open spaces, they provide shelter for our birds and insects as well as the fruits and nuts necessary to sustain much of our wildlife. Trees can block wind on a particularly open site or maybe provide some much needed shade. They play an important part in countering the greenhouse effect by absorbing many of the pollutants and moisture resulting from the carbon producing age we live in. But for me trees are just beautiful.

Trees, of course, can often grow far too tall for our gardens but there are dwarf species – cultivars that can safely be grown in your garden, but make sure you do choose the correct cultivar or you and your neighbours might get a surprise you don't want, especially when trying to remove the roots!!

Plant growers have over the years also produced many smaller, ornamental tree cultivars, which can successfully be planted in pots and containers, but don't forget to water them frequently.

So I hope I have clarified genus and species and family for you, but I am sure that you have all noticed when buying a plant that there is usually a third name on the label which describes the variations that can occur. Let me briefly explain. Firstly, and most commonly is the cultivar.

A cultivar is a variety of plant that is especially grown by plant breeders whereas plant species occur naturally. A cultivar name comes after the genus and species names e.g. *Chamaecyparis lawsoniana* "Elwoodii" – what a mouthful!

When purchasing a new plant make sure you choose the right cultivar – look at the label. Size is important! When ordering a tree for example not only eventual height but also width must to be taken into account. Select the appropriate species and cultivar as a tree more than 16 feet might be too large for smaller gardens and some trees, of course, grow well over 100 feet!

I remember the day my wife bought a small eucalyptus in a pot because she liked the leaves. She planted it and it grew and grew! The same of course also applies to shrubs – ensure you have enough room for them to grow or drastic pruning will be required on a regular basis

Cultivars can only be reproduced vegetatively as they will then be identical. They may lose the cultivar identity if they reproduce by pollination (two parents) as they will not usually come true to the parent. Where you see Var (variant) - the appearance of the plant may be different and distinct from those grown under differing conditions or region. One particular example is *Cornus kousa* var chinensis. This particular variant of Cornus originated in China.

The term Forma may be used to highlight the difference between a plant species grown high up a mountain and another at the bottom of the valley. The same species of plant can be grown in two very different environments - one alpine, one a moist shady setting.

Sports are the name given to genetic abnormalities often resulting in some strange looking plants. Some woody or some herbaceous plants may be different in colour, foliage or simply shape. These plant sports may be propagated to retain the sport characteristic if it is sufficiently interesting or visually different.

So many plants, so many names to remember, but what can we do to help the survival of these families which are integral to the survival and well being of the human race.

There are some excellent books available featuring plant names.

William T. Stearn – Botanical Latin (Reference Book)

Lorraine Harrison – RHS Latin for Gardeners (Illustrated)

Martin Page - Name that Plant (Quick Reference)

The RHS Lindley Library in London, UK is a botanical treasure house of printed books on gardening, botanical art and photographs. It also holds archives of notable gardeners and designers. It is the largest horticultural library in the world and also has a wide range of reference books and botanical material.

I wish I could find more time to spend at this Library, but the grass needs cutting, that shrub must be pruned and the weeds over in the back border definitely need some attention.

Check your soil before you plant.

Chapter 7 Digging for Victory or the importance of soil

I have tried to stress to you some of the many benefits that we the human race get from plants. The truth is we could not live without them.

Now it's pay-back time. What do plants ask for in return? For some not a great deal. Many plants seem to appear from nowhere, spread by seed blown in the wind, or transported by the droppings of birds or small mammals casually roaming in our gardens and green spaces. The wild flowers that set up home in our highways and byways. Those dreaded weeds that always seem to turn up, unannounced and mostly unwelcome. These plants naturally thrive.

But (and I know I shouldn't start a sentence with but) but - in the main if we want our farmers to produce successful crops, or if we want our gardens to be green, lush and full of happy plants, then it makes sense to provide the best possible conditions so ensuring strong healthy growth in turn yielding as high a quality crop as possible and the most attractive and long-lasting blooms from our ornamental plants.

Here is where a little scientific knowledge can be very beneficial to growers and gardeners alike.

To get the most out of your plants first you need to understand your soil. My young grandchildren have not yet come to understand the word 'soil'. They think that I grow my plants in 'dirt'. It is a little more complicated than that!

When working your plot does your trowel or spade make a terrible 'clunk' noise, and leave you aching and reaching for the Radox!

If this is the case some hard work is required before you can plant successfully. Dig your soil over well and to try to remove as many large stones and old roots as possible. Add some topsoil (ok you're thinking, so that's the stuff that goes on top) or loam (what's that?) and organic matter (where does it come from?) to help break up the soil. You'll probably still need the Radox though!

Take a look at agricultural fields and see how well the soil is prepared in advance, with ploughing, fertilising and careful seed sowing and planting. Loam, sometimes referred to as the texture of soil, is a combination of clay, sand and silt. In some geographic areas chalk will also make an appearance, but of course all these constituent parts are naturally found in hugely varying proportions. Healthy soil must also contain air, water, minerals, organic matter and, vitally, living creatures such as microorganisms and worms.

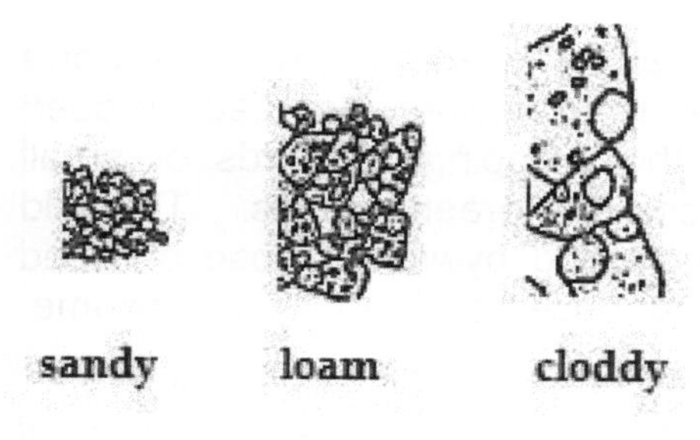

Clay particles are small and clump together, maybe believing there to be safety in numbers, and are thus able to absorb more water along with the nutrients it contains. Alas once water supplies begin to dry up clay particles are left clinging even tighter together, making the ground difficult to dig and resulting in a compacted soil.

It has to be said though that clay particles have a special property in that they are covered in tiny electrical charges called cations and these can be useful in holding onto nutrients rather than allowing them to be washed or drained away by all but the heaviest rain storms or floods.

Silt particles are also small and therefore able to retain moisture.

Not many nutrients in this soil

Sandy particles, on the other hand, are larger and more stand alone, but unfortunately the space around them allows water to drain through and essential nutrients are lost.

Chalk drains so well that it is sometimes called porous.

The aggregate parts of loam will therefore contain particles of differing matter of differing sizes, all held together by a rotted organic sticky black matter called humus which is formed by the decomposition of leaves and other plant material by soil microorganisms.

So you will understand that it is the variation in particle size of the constituent parts of loam that affects their surface area and therefore the amount of water and nutrients that can be retained in the soil.

Drainage of water is affected by gravity, loss of water through the plant by transpiration and evaporation of moisture from the soil surface.

As we all know, soil needs to be well broken up, commonly called 'tilling' which is important for good drainage as well as allowing for root and air penetration.

In growing media, a fancy name for the composts we buy to add to our gardens, a mixture of 30% clay, 40% sand and 30% silt is a recommended average loam balance for water retention and drainage, but that's only part of the good soil story. Manufactured compost formations usually make sure the loam is sterilised to eliminate any weed seeds and pests and diseases.

What does your soil consist of? Try one or more of these simple tests to help you find out.

Perhaps the easiest way to analyse soil is by touch and if you have some youngsters around, you may find you have some willing hands to help in the sausage making.

Take a small handful of soil and try rolling it into a sausage shape. If you can and it has a sticky feel, it's clay. If it is only just holding its shape and feels somewhat soapy, it's a silty clay. If it just falls apart it's sandy, one of the hardest soils to cultivate and may need a system of irrigation to be implemented as water will be constantly draining through.

Why Kew Gardens was moved to a site on sandy soil is a mystery to me. Perhaps they wanted a challenge! Perhaps no-one did a soil test!! So remember - clay feels sticky when wet, has good water and nutrient retention but poor drainage.

Silt feels silky and soapy when wet and drainage can be variable depending on the composition of the silt which comes from river estuaries. Silt can have a high nutrient and mineral value and is one of the most fertile of soils.

Sandy soils feel very gritty, are fast draining and do not hold water or nutrients.

Another test is to place a sample of soil from the location to be tested into a clean dry jar, add some de-ionised (distilled) water and shake thoroughly. Leave a while to settle. You will find the soil forms separate layers. Perhaps, as you would expect, any larger stones, heavier sand or grit will fall to the bottom of the jar. The middle level will contain very small stones, small grit and light sand. Sitting on the top will be any organic matter, small light clay and silt particles.

Do you have that ideal mix of 30% clay, 40% sand and 30% silt or do you need to add some soil improvers such as leaf mould, compost or manure?

How moisture retentive is your soil? Try pouring water on a section of soil in your flowerbed and time how long the water takes to disappear below soil level. If the water is still sitting on the surface after one hour, the soil is clearly very poor draining and may be seriously compacted clay and contain a large amount of builders rubble, especially the case if you are creating a garden from scratch at a newly built property. If the water disappears in under a minute drainage would appear to be satisfactory. If the water disappears instantly then it may indicate a sandy or chalky soil

Chalky soil is alkaline, very free draining, holds little water and like clay soil can dry out easily. Chalky soils can be fertile, but many of the important nutrients especially iron and manganese required by ericaceous plants are blocked as the pH is so high, due to the high amounts of calcium being present. A bit of chemistry for you!

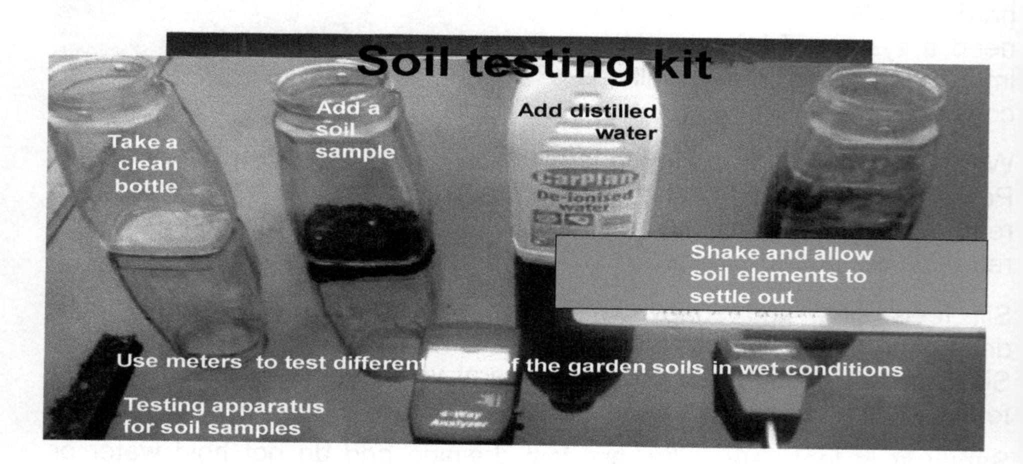

A pH measurement is an indicator of how acidic, neutral or alkaline the soil is.

Electronic soil test meters are reasonably priced, invaluable aids for gardeners to easily and quickly check pH, soil moisture and fertility (nutrients).

Wipe the meter probes with a clean dry cloth and select the appropriate setting for pH, fertility or moisture.

Insert meter probes into several different areas of preferably wet or moist soil making a note of the readings. pH readings must be taken from moist soil.

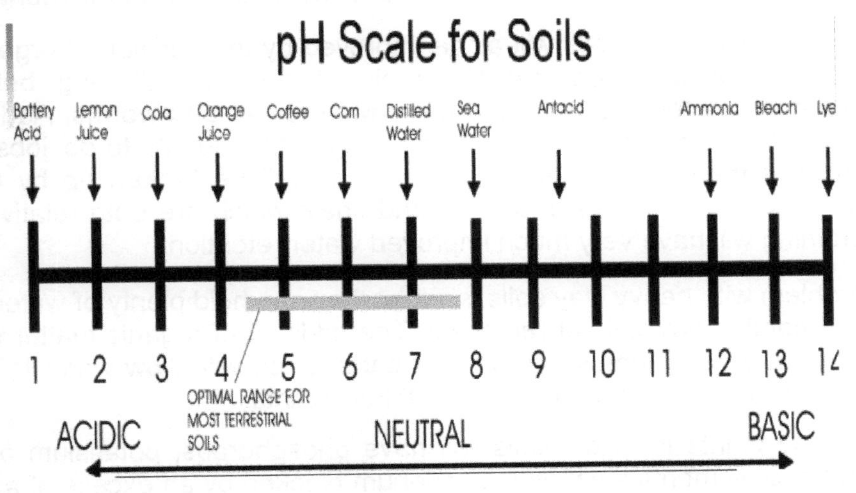

Acid loving plants prefer to grow in the pH range 4.5 – 6.5. Lime loving plants prefer a pH range of 7.5 – 8.5

When you are spending a happy few hours wandering around in your local garden centre please do not be tempted to buy a plant just because it looks good. Always check the plant label to see what soil is recommended as you may need a particular soil pH for that plant to thrive. Having said that, most plants are quite happy being grown in neutral soil although some are a little more choosy and have a preference for a particular soil or growing conditions. Just as well really, different soils have provided us with a far greater diversity of plants.

Plants that like acidic soil are called 'ericaceous' and plants that like alkaline soil are called 'calcareous', although it is easier to say 'acid loving' or 'lime loving'. Ericaceous plants in particular will not thrive unless grown in specially formulated growing media and compost as the clay based formulations, produced for neutral and alkaline plants, contain calcium which will block out the nutrients of iron and manganese which ericaceous plants particularly depend upon.

Ericaceous formulations appear to be a simple combination of acidic organic matter such as peat, coir, bracken, pine bark or leaf mould and instead of clay there appears to be added grit and sandy loam. As with baked beans, different companies produce their own variety of tried and tested formulations and over time a gardener will find his or her favourite.

Clay, sandy and chalk soils will all be improved by the addition of organic matter, manure and compost to break up the soil allowing better penetration of water and providing a pathway for air and worms. Adding some organic matter to your flowerbeds is one of the 'ought to do' jobs in autumn and then again in early spring. Clay will be broken up by the organic matter over winter and sand and chalk which are both relatively free draining will have very much improved water retention.

The problem with heavy clay soils is whilst they can hold plenty of water in warmer months they will not release it. The addition of organic matter will break the large clay lumps into smaller parts which will allow water to be released more easily when required by thirsty plants.

Lime-loving plants in acidic soils will have phosphorous, potassium and certain trace elements such as molybdenum blocked by an excess of acid hydrogen ions causing the acidity in the soil. These lime-loving plants will not progress without these minerals and nutrients.

To solve the problem, extra garden lime made from pulverised limestone or chalk can be added. Lime is also available in the form of small pellets which are easier to apply.

Over time you may need to add some calcium as soil gradually becomes more acidic because of bacterial action constantly taking place in the soil and the acidity from rain.

To alter soil towards alkalinity, add calcium carbonate, ground limestone, dolomite (magnesium) or calcified seaweed. A mushroom based compost is also highly recommended.

Vegetables generally prefer limey soils. As you know, there always has to be an exception to the rule, and carrots are known to grow satisfactorily in soil pH of between 6 and 7, which is slightly acidic. Brassica vegetables however can suffer club root disease at this pH and must have alkaline conditions between 7 and 8.

Pinks, Viburnum, Delphiniums, Hellebores, Rudbeckia and Clematis are also all examples of lime-loving plants. Plenty of colour there to brighten up your garden.

Technically it is possible to attempt to change a soil's pH with chemicals but this is very difficult, expensive, time consuming and is not recommended because you may well end up with only a mixture of soil pH. Your plants will be very confused and unhappy, especially those with a specific preference towards acidic or alkaline soils.

When buying compost or soil do take time to look at the product description on the bag. Read it carefully and it will help to avoid costly mistakes.

Acid loving plants such as Rhododendron, Magnolia, Azalea and Heather will not thrive in an alkaline soil. The elements of iron and manganese which are blocked by calcium in lime-based chalky soil are vital for ericaceous growth. To give your acid lovers a boost there is certain action you can take, but you must decide which is best for you.

Do you remember your chemistry lessons?

Group→ ↓Period	1	2	3	4	5	6	7	8	9	10	11	12	13	14	15	16	17	18
1	1 H																	2 He
2	3 Li	4 Be											5 B	6 C	7 N	8 O	9 F	10 Ne
3	11 Na	12 Mg											13 Al	14 Si	15 P	16 S	17 Cl	18 Ar
4	19 K	20 Ca	21 Sc	22 Ti	23 V	24 Cr	25 Mn	26 Fe	27 Co	28 Ni	29 Cu	30 Zn	31 Ga	32 Ge	33 As	34 Se	35 Br	36 Kr
5	37 Rb	38 Sr	39 Y	40 Zr	41 Nb	42 Mo	43 Tc	44 Ru	45 Rh	46 Pd	47 Ag	48 Cd	49 In	50 Sn	51 Sb	52 Te	53 I	54 Xe
6	55 Cs	56 Ba		72 Hf	73 Ta	74 W	75 Re	76 Os	77 Ir	78 Pt	79 Au	80 Hg	81 Tl	82 Pb	83 Bi	84 Po	85 At	86 Rn
7	87 Fr	88 Ra		104 Rf	105 Db	106 Sg	107 Bh	108 Hs	109 Mt	110 Ds	111 Rg	112 Cn	113 Uut	114 Fl	115 Uup	116 Lv	117 Uus	118 Uuo

I have found the ideal combination is adding slow release ericaceous granules over winter and adding ericaceous liquid feed (sometimes known as sequestered iron which includes added Manganese (Mn) and Magnesium (Mg)) during flowering time, whatever the season.

Adding some ericaceous soil and organic mulch will also provide much needed nutrients over the winter. This appears to have had a remarkable effect on the deciduous Azaleas in my garden.

In addition it is also possible to improve soil acidity by adding ericaceous organic matter, peat, bark or pine needles. On a larger scale it may well be easier to use Iron sulphate or Ammonium sulphate but most effective is adding horticultural sulphur.

Did you know earthworms dislike acidic soil, so you won't find any in your ericaceous beds!!

What can we do to help nourish and nurture our plants?

A healthy plant needs a combination of five major nutrients to keep it in good condition, much as humans need certain vitamins and trace elements.

N - Nitrogen helps to keep plants green
P - Phosphorus is important in cell division and strengthens developing new tissue
K - Potassium and Mg – Magnesium both aid flower and fruit production
Ca – Calcium strengthens cell walls and promotes vegetable growth

All the above elements are vital for plant growth but are required in very different proportions. Firstly, they are needed to build the cells, the tissues and the organs of a plant.

Secondly, they are used for the chemical and hormonal balance of the plant helping to produce enzymes. Plant enzymes are very important to boost plant growth.

The third and really important function of the nutrients is to maintain the chemical balance of the plant, so allowing osmosis, which is the movement of water through the plant membrane. Osmosis maintains the turgidity (water pressure) within the plant keeping it upright.

Other important trace elements required are Manganese, Iron, Boron, Molybdenum and Zinc. Deficiencies in these elements can cause growth and plant health problems especially if plants are in the wrong soil as we have just chatted about. It's a good idea if you are having problems to take soil tests for fertility as there may be a shortage of specific nutrients.

Trace elements are important for plant colour and flavour in edible vegetable plants.

Extremes of pH such as soil found under pine trees (pH 2 – 4) or very chalky alkaline soils (pH 8 – 12) will upset the N-P-K-Ca-Mg nutrient balance and can lead to difficult plant growing conditions.

There must be a careful balance of fertiliser applied to plants or this too could lead to an imbalance in these essential minerals, reversing the vital flow of water through osmosis back into the soil. Too much nitrogen and you will find your plant becomes all leaves and no flowers. Too much potassium will block nitrogen, turning leaves yellow and affecting photosynthesis.

Fertiliser products can have a huge effect on specific soil requirements that gardeners need to evaluate carefully. Considerable expense can be involved in composting, manuring and fertilising so spend time taking or sending off for fertility and soil measurement tests before you order fertiliser products and if necessary take advice!!

In fact, your soil may not require fertilising, just a good soil/organic matter balance to keep the roots happy. Some gardeners add mycorrhizal fungi to plant roots which helps bind them to the soil.

It is very important for the fertility of soil to have organic matter in it, especially as it contains billions of microbes to help produce nutrients, one particularly important being nitrogen.

Constantly removing nutrients from soil makes it more difficult for plants to grow as time goes on. Farmers try to overcome removal of nutrients by leaving fields fallow for a year or two and may add legumes such as clover that have the nitrogen nodules particular to the legume plant family. This has a huge effect on renewing nutrients.

Never add lime and fertiliser at the same time; the two products cancel each other out and your soil will become poorer for it. I have to say the resultant release of ammonia as a gas is not at all pleasant!

Organic fertilisers from natural sources are made from plant and animal waste and should be used to increase soil fertility, but are slow acting and depend on temperature and moisture content. Organic matter is derived from anything that is or was living and acts as a vital soil improver by breaking up soil clumps, providing air and water vents that allow roots to move freely to access water, nutrients and oxygen for respiration of the roots. Let's not forget it helps worms move and worms help break up hard clay soil.

There are three main types of organic matter and a ratio of approximately one part of the chosen organic matter to ten parts garden soil is recommended.

Ready to spread compost

Compost, decomposing vegetable waste which is mainly used as a soil improver.

Make sure manure is well rotted

Manure, decomposing animal waste which is more nutritious in nitrogen than compost and also helps to break up hard clay soil.

My grandson lends a helping hand

Leaf mould – is one of the best soil improvers. Make your own by collecting leaves and placing them in bags with holes to allow in air and moisture to help the leaves rot down.

If you are lucky enough to have a compost heap in your garden, treat it kindly. Only add vegetative matter, protein food stuffs attract vermin, and don't add thick woody stems as they take too long to rot down.

Turn the heap regularly, keep it moist and it is highly recommended to add some straw. Once rotted, add this all important compost to your garden. Your garden waste will certainly not be wasted.

Compost and manure decompose to create the perfect planting environment for you. The microorganisms, (bacteria and fungi) found in soil, together with earthworms create a dark substance called humus which is rich in nutrients, vital for soil improvement and important for water retention.

Mycorrhizal fungi in soil form a two-way network association between the fungus itself and the roots of its host plant. The fungus exists by taking sugars from the plant in exchange for moisture and nutrients from the surrounding soil, so creating a more vigorous root system and a healthier plant.

This fungus is particularly effective in organic and non dig cultivation where fertilisers are not used. It also works well in soils lacking phosphorus alleviating the necessity to add huge amounts of fertiliser which would otherwise be required. A good example of nature coming to the rescue!

useful chaps, worms

Earthworms are helpful creating passageways through the soil, improving aeration and drainage. Worms also drag leaves into the soil for processing by bacterial micro organisms; at the same time they eat the soil producing finer porous particles.

The inorganic loam mixture of clay, sand and silt together with the organic humus creates an ideal soil which is crumbly and well drained with porous air holes. These very important air holes allow plant roots to spread through the soil and absorb moisture, nutrients and oxygen from the air. It's also very important for your plant roots' respiration (burning energy) to be able to get rid of carbon dioxide back into the air. Heavy compacted soil can cause death to roots by carbon dioxide poisoning - too much water causes death by drowning – not good for the plant and upsetting for the gardener.

There are many inorganic chemically manufactured products on the market supplied in solid (granular) or in liquid form used for a wide range of specific soil requirements such as for adding more Nitrogen, Phosphorus, Potassium, Calcium, Magnesium and Sulphur and minor elements which can be applied direct to the soil as more concentrated nutrients.

Growmore NPK 7.7.7 invented in wartime was very successful in boosting food and vegetable production. It is water soluble which is vital for a folia feed, where leaves absorb the required nutrients. Growmore dissolves quickly in soil.

Deficiency in nitrogen is shown by yellowing of leaves (chlorosis) with red tinges.

High nitrogen fertilisers such as ammonium sulphate 21%N or nitrate 35%N are very effective as they are soluble and boost green leafy growth.

Deficiency in phosphorus is shown by stunted growth and very dark green leaves with veins turning purple.

Phosphorus does not move around in soil as is the case with some other nutrients, as it is not particularly soluble. Phosphorus is required in smaller quantities compared to Nitrogen, but can be deficient in soils with little organic matter or with an acidic soil pH of less than 6.

Phosphorus fertilisers for stimulating root and shoot growth are supplied as single super phosphate with an average 20% phosphate or triple super phosphate with an average 48%.

Deficiency in potassium is shown by leaves going blue green with the edges turning brown. Potassium fertilisers will be required for nutrient poor sandy soils lacking in clay content. Used commercially to boost flowering and fruiting, potash is supplied in a highly concentrated form (up to 60%) as potash muriate for outside agricultural crops, as a potassium sulphate (up to 48%) for crop growing under glass. For soluble folia (leaf) liquid feeds they are supplied as potassium nitrate which is 46% potassium and 13% nitrogen.

Deficiency in magnesium is shown by leaves going yellow between the veins.

Magnesium also comes from clay and again may be deficient in non-clay sandy soil. It is a vital component of photosynthesis.

Deficiency in calcium is shown by brown colouration in plant tissue and young leaves turning yellow.

Most soils have plenty of calcium, except in sandy or acidic soils. It is vital in cementing cell walls together and producing cytoplasm in the plant cell. Apples and tomatoes may need added calcium as well as lime soils.

Sulphur is contained in most inorganic soils and in rain water.

If you prefer a less chemically based approach then the following may be for you.

Bonemeal products contain a ratio of 5 parts nitrogen to 20 parts phosphorus, especially formulated to aid root growth.

Hoof and horn is used in the base formula of John Innes fertilisers and is a similar product to dried blood in nitrogen content. As hoof and horn is slower to release nitrogen it is preferred for organic crop growing because there is less chance of leaf burn damage.

Fish Blood and Bone is a well-balanced organic fertiliser. A more natural product it is popular with organic growers despite having reduced levels of phosphorus. Usefully it can be purchased in a quick release form.

Urea is a quick acting foliar feed with a very high (46%) nitrogen content particularly used by the agricultural industry.

Don't be afraid to take advice as to the best product to buy from a horticulturist or at garden centre. To help them to help you, take a leaf or stem from the affected plant for them to look at. Also do a soil test preferably with a simple electrode tester and make a note of your findings, it will help with the diagnosis. You can also send soil samples from different parts of your garden to advertised laboratories who will undertake a full and specialist test on your behalf

Many nutrients are recycled and of great importance is the nitrogen cycle. I will try to explain this simply.

Air is 79% nitrogen. Living things need nitrogen to create the proteins necessary for survival. A plant of course is a living thing, but how can nitrogen in the air help it to grow? It needs the help of some unseen workers.

Soil and organic matter contain untold billions of microbes, bacteria and fungi, the lifeblood of this planet which work away continuously (unpaid and uncomplaining) digesting and decomposing.

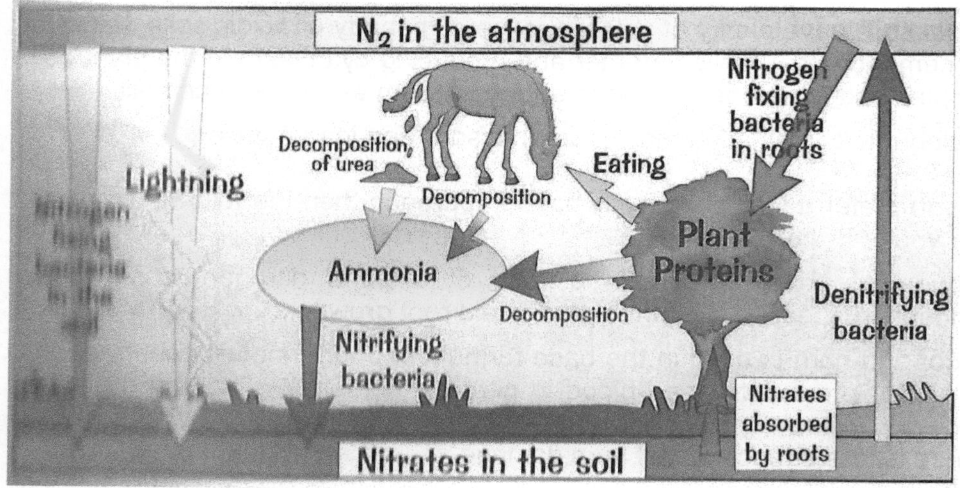

(picture copyright CGP books Ltd)

Nitrogen is taken from the air by these tiny single-cell micro organisms into the surrounding soil. Because nitrogen in gas form cannot be absorbed by plants the hardworking microbes capture the gas turning it into ammonia and ammonium compounds called nitrites and nitrates. This process has been called ‘nitrogen fixing’.

Perhaps the easiest way to explain this is that these microbes take up the nitrogen, and then excrete it again as nitrates which are then broken down into nitrites and ammonium ions - just what a plant needs! Moisture from the soil containing these soluble chemicals is sucked up into the plant via its roots and transported through the xylem vascular vessels to all parts of the plant.

The plant then uses these chemicals to make useful products including the vital chlorophyll which is made by photosynthesis. They are also used to make valuable proteins including amino acids which are required for building the plant cell structure, its tissue, organs and DNA. DNA is vital in producing the plant cell nuclei which is the control panel in plant cells.

To complete the cycle, nitrogen is eventually returned to the atmosphere from the soil in the form of a gas. Nitrogen gas is produced by special de-nitrifying bacteria in the soil making more nitrites which release the gas back into the atmosphere. Decaying manure and compost assist this process.

Another way to introduce nitrogen into the soil is by the addition of animal waste in manure. This is very beneficial as its nitrogen is absorbed directly into the soil, but should only be used in a garden when it is well rotted or it will badly burn the leaves and roots of the plant because of the strength of the ammonia during the initial rotting process.

A third way of artificially fixing nitrogen from the air was invented by two German scientists during the First World War to provide a source of ammonia which could be used in the production of explosives. The ammonia produced by this process (known as the Haber process) is still used today in the manufacture of fertilisers.

Some clever plants such as the Pea Family (*Leguminaceae/Fabaceae*) also capture and store nitrogen in root nodules so avoiding the necessity of adding extra fertiliser.

There are of course many soils, composts and manures that have been developed and marketed as soil improvers. One example is peat which for a long time was considered one of the best moisture retentive additives a gardener could use.

Unfortunately, a major problem with peat extraction, as with many natural resources, is that it cannot be replaced.

use peat sparingly

Peat bogs are made up of partly decomposed vegetative matter sitting on boggy acidic ground. Because it is a naturally decayed material peat is very good at holding at lot of moisture enabling nutrients to move around easily in the surrounding soil.

These bogs are environmentally friendly helping to prevent flooding, and acting as a carbon sponge better, it is said, at preventing harmful carbon dioxide escaping into the atmosphere than many forests.

Sphagnum moss is the living plant which lies on top of peat bogs. It can absorb vast quantities of water, adding hydrogen ions to the moisture which is why it has been found to be such an ideal acidic addition to ericaceous soils and composts

I understand that sphagnum moss is a popular plant used by the carnivorous plant growing industry as it has such a high moisture content but absolutely no minerals which strangely kill off a 'meat eating' plant.

Peat it must be emphasised is a natural resource and must be used very sparingly, not as a general mulch. Sadly 95% of the UK's peat bogs have disappeared, leaving a black emptiness in their wake. What took millions of years to form took just 40 years to deplete. There are further peat sources to be found, but worldwide environmental concerns regarding the depletion of these sites have prompted many countries, Australia in particular, to rescind the required extraction licences.

We must live and learn, but in this case I fear it is too late - but perhaps not quite. There are peat free alternatives on the market. We must look for them.

There are some impressive efforts taking place in areas of the UK where attempts are being made to regenerate areas of sphagnum bog by incorporating mass plantings of sphagnum plug plants along with added hormone stimulants. This project was covered recently by the UK's BBC's Countryfile Programme, which I must say is my favourite Sunday evening viewing and a source of much useful information.

I personally have found John Innes soil improvers excellent for growing plants in containers because of the water retentive loam very much required for larger plants, particularly those being potted on.

Again, look at the bag to check whether you are buying the correct product for the job you have in mind. Do you need a formula for seeds, to pot on young seedlings or to cope with the requirements of a fully grown plant? Research undertaken has seen the development of specific formulations of soil for different stages of a plant's growth.

John Innes was a 19thcentury land and property dealer, who very generously left his fortune to be used for research into the improvement of horticulture. A man with a green vision. I am sure he would be proud of the research undertaken in his name. The John Innes brand is recognised and respected worldwide.

It was research undertaken by the John Innes Institute that discovered a heat sterilisation process which eliminated weeds, pests and diseases from their composts, so you can be sure you are giving your plants the very best start in life

Formulations for neutral and alkaline growing start with the compost recommended for sowing seeds and cuttings, generally comprising two parts sterilised loam soil, one part peat or peat substitute and one part coarse sand. You can't give a baby steak and chips and seeds will not require an added fertiliser. As plants become bigger and more established they can cope with an increasingly diverse diet.

Potting-on composts in general have more than double the soil content with added peat or its substitute, sand, and some fertilisers of super phosphate and potassium to boost plant growth. There is a base mix of fertiliser comprising hoof and horn which is high in the nitrogen necessary for greening a plant's cells.

Following on from seed and potting compost, the next three formulations, John Innes 1, 2 and 3, add proportionately more sterilised loam, have the same fertiliser base but also some added chalk to raise the pH level for a more alkaline soil.

Gardeners should feel free to use these formulations and in fact mixing instructions are available on websites, although a visit to the local garden centre to pick up supplies would seem a far easier option.

Having looked carefully at formulations used for ericaceous compost and from information received from the John Innes Manufacturers Association, the loam content of ericaceous soil is very much sandy topsoil with a reduction in clay and calcium containing loam. The proportion of peat is increased slightly to compensate for the lower organic content in clay loam. This change in calcium content formulation enables the pH in the soil content to be much reduced for the whole mixture.

Currently more trials are taking place into the formulation required for ericaceous soils, incorporating excellent water retaining materials such as ground charcoal, ground waste wood and coir in place of the peat based materials which are being restricted by the Government's environmental agencies.

It will be interesting to see whether the high peat content of 50- 60% in some ericaceous formulations will eventually be reduced and replaced with peat substitutes

Chapter 8 Growth and Reproduction

So, you have provided the perfect growing conditions for your plant. The soil has been dug and you have a perfect loam raked and raked yet again to produce a fine tilth – no lumps (have you ever made pastry - you must get rid of the lumps). What more could a plant possibly want?

Well – food, nutrients and water for a start. Like all living matter plants must have food and water to survive.

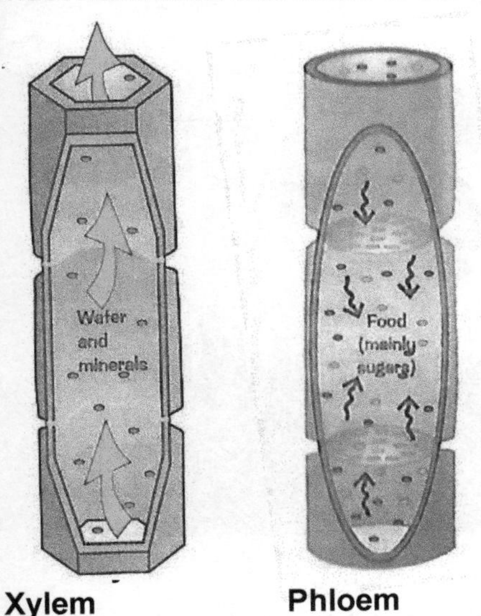

Xylem **Phloem**

Osmosis is the movement of water through the vascular system (think veins) of a plant. It is the water in a plant's cells that keeps it upright and rigid (turgid).

(picture copyright CGP books Ltd)

Water dissolves the minerals found in soil. This nutritious combination of water along with the minerals, in the form of tiny charged particles called ions, is sucked up through a plant's roots and transported by the vascular system through the xylem vessels to all its living cells.

So the NPK and Ca and other minor elements I mentioned earlier all enter the plant through its roots by a combination of capillary action and osmosis. A continuous movement of sucking up and pumping round.

Eventually the water reaches the leaves where it evaporates through little holes called stomata. This process, called transpiration, signals the roots to suck up more water. More water. More nutrients. More growth.

This all important movement of water through a plant depends on rainfall, light levels, humidity, temperature and wind although the extremes of any of these can destroy a plant. That, of course, is only half the story.

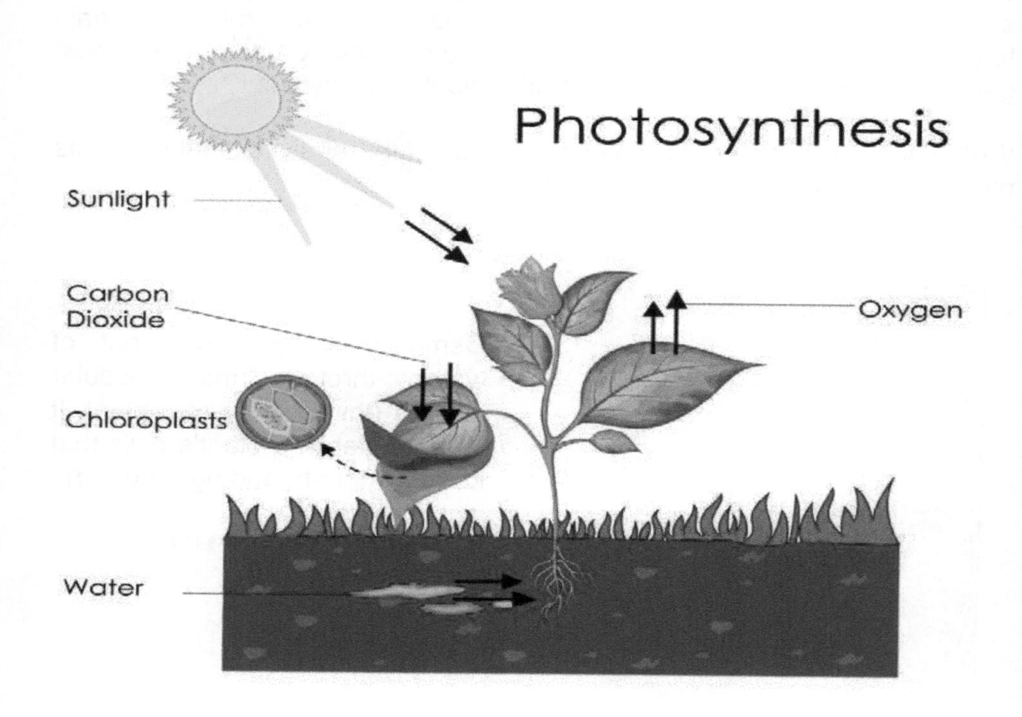

Plants cannot live by water alone. They also need to produce their own carbohydrate food, the simple sugars of glucose and sucrose. These sugars, formed by the process of photosynthesis, are transported to all parts of the plant through the vascular phloem vessels.

Photosynthesis takes place in the energy producing chloroplasts mostly found in the leaves. Chloroplasts contain molecules of the green pigment chlorophyll. Chlorophyll (which comes from the Greek for leaf and green) converts the light energy of the sun into the sugars needed by the plant cells for healthy growth.

This carbohydrate food (glucose sugars) is made by combining sunlight, carbon dioxide from the atmosphere and water which is then transported around the plant through the vascular vessels (phloem) to all parts of the plant, including the roots.

The food is then used up in the reverse process called Respiration, when oxygen is released back into the atmosphere.

Respiration uses oxygen from the air to burn the carbohydrate (sugars) produced by photosynthesis to provide the energy for all the plant cells to multiply and grow. Carbon dioxide is released back into the atmosphere.

This is shown simply by the following two cycle processes which depend on light intensity, temperature, water, carbon dioxide and oxygen.

Plant Photosynthesis
CO_2 + H_2O + Sunlight = Carbohydrate + O_2

Plant Respiration
O_2 + Carbohydrate = H_2O + CO_2 + Energy

These two simple processes keep human and animal life fed by growing all the plants on the planet!!

Remember - xylem vessels move water, nutrients and minerals from the roots; phloem vessels move food and minerals from the leaves.

The more efficiently balanced the movement of food and water the faster the plant cells grow into the tissue required for new roots, leaves and flower organs.

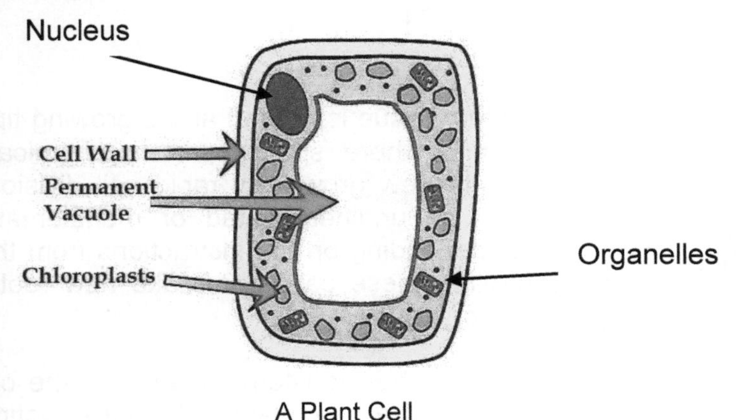

A Plant Cell

(picture copyright CGP books Ltd.)

A stem cell is the basic building block of living matter, but one that has yet to find its purpose in life – it needs to be told what to do.

Plant stem cells receive their instructions from the nucleus, the large black dot in the above diagram, (think of it as the cell control panel) where the controller is in charge of directing the growth operation. All plant cells contain structures named organelles which are designed to allow specific functions to be carried out. One very important example of an organelle is the chloroplast which is essential for the process of photosynthesis to take place.

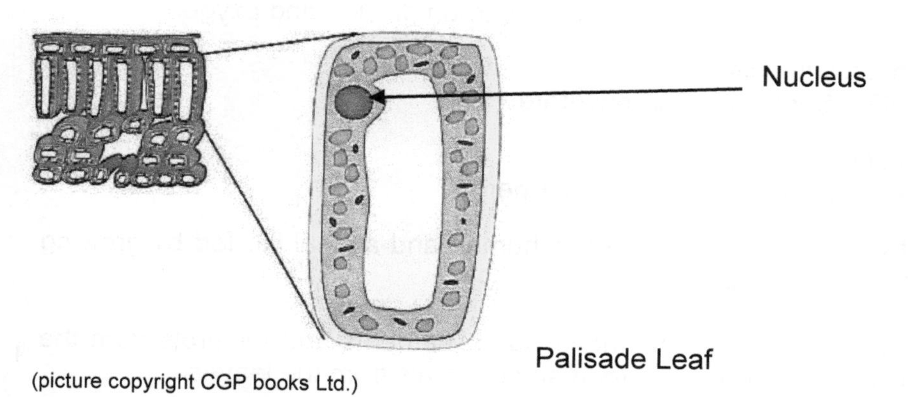

(picture copyright CGP books Ltd.)

The above diagram shows the cross section of a typical leaf. The main function of a cell is the absorption of light by the small green chloroplasts contained within the cell wall so that photosynthesis can take place to produce the glucose food energy for the plant.

Budding meristem tissue

Meristem tissue is formed at the growing tips of plants where special protein chemicals stimulate new growth by rapid cell division. Known as undifferentiated or meristematic cells, depending on the instructions from the controller, these cells can make new roots, stems or leaves.

When a gardener takes a cutting it is the meristem tissue from the cut material that will form the beginnings of a new plant. Hormone rooting powder can be used to help give the new plant a good start in life.

The African Violet, Begonia and Sedum are good examples of plants that can be propagated simply by leaf cuttings. The meristem tissue from cut leaves will quite quickly send out roots. The controller nucleus of the root cells then directs the meristem tissue to produce a stem, leaves and finally the flower, a lovely new plant!

Gardeners may be puzzled that plants do not always behave according to plan. For example plants may not flower when we think they should, and then unexpectedly bloom several years later. The Liriodendron tree may take up to 25 years to produce its first flower. The Madagascan Palm, another late starter, may not flower for 100 years. Hormones may be involved here as well as biochemical imbalances that cause what is known as extended juvenility.

We all know that humans have hormones (although sometimes it seems teenagers have more than their fair share) but interestingly so do plants.

Humans have hormones in their blood, plant hormones are vital chemicals found in sap. They send messages to all plant cells which trigger how the plant should behave. Whether a plant turns towards or away from light, whether a flower opens when the sun shines and closes at nightfall, whether it sends out that enticing scent during the day or waits until evening, it is all down to the messages delivered by the hormones.

Hormones dictate shape, rate of growth, when a flower should appear, when seeds should be produced and when and if a plant should die. Somewhat 'bossy' perhaps, but a plant would be nothing and could not survive without its hormones – nor, of course, could we.

Auxins are hormones which help elongate plant cells by speeding up cell division – in other words they assist growth. Gibberellin is another hormone encouraging stem growth; a lack of this hormone causes plants to be 'dwarf', but add more and a plant can become a 'giant'. Perhaps this is what happened to Jack's beanstalk.

Synthetic auxins produced commercially are used not only for the very important rooting powder which encourages new growth in cuttings but also in hormone based weed-killers where weeds will be made to grow too quickly and then die back.

For those agriculturists not involved in producing organic crops selective weed killers have been produced which will only target weeds, leaving the crops and lawn grass unaffected. There are also selective weed killers for use on lawns, targeting dandelions and other unwanted weeds, letting grass retain its lush greenness. Do not be tempted to use a granular weed killer on grass unless you can be sure that rain is imminent otherwise you will find that not only have the weeds disappeared, but so has much of the grass, leaving some very unsightly brown patches. Perhaps those dandelions weren't so bad after all!

Cytokinins are produced in the roots of plants and travel up the stem through xylem vessels. They help with cell division, seed germination and the speeding up of buds to flower. They also work alongside auxins to increase the green chloroplasts that enable photosynthesis to take place.

Seeds will often remain dormant with hormone chemicals called inhibitors holding up germination until conditions are right for growth to resume, then in come the Cytokinins to help get things started.

Absissic acid is a hormone produced by the plant in winter which assists in the ageing process by reducing photosynthesis, weakening the cells and eventually causing leaves to drop.

An addition of seaweed is excellent for plants as it is a source of hormones and very important trace elements (containing over 50 minerals) although there is little NPK nutrient value in seaweed and it appears to be especially lacking in phosphorus.

Some hormones are used in natural genetic engineering to boost plant fruiting, size, strength and growth. Ethylene gas is used commercially to induce flowering and ripen fruit. This is achieved by stimulating the creation of more chromosomes per cell of the fruit, known as polyploidy. This is very important 'natural' genetic engineering but is also something plants may themselves do naturally, covered in Chapter 12.

As we all know, some plants appear in the spring, many in the summer but some plants wait for autumn before they put in an appearance. More often than not it is light which activates the hormones which in turn send out the 'wake up and get growing' message rather than temperature.

Photoperiodism is the name given to the reaction of plants to the amount of daylight they are subjected to. Strictly it is the amount of darkness a plant is exposed to that counts. Those plants not affected by the light/dark balance are known as 'day neutral'. Some plants, such as chrysanthemums and poinsettias, come under the heading of short day plants as they like less light and are therefore happy to grow in the shorter days of the autumn. Other plants need the longer daylight hours of summertime to thrive

Plants often grow faster as the temperature increases but some need a cold period called vernalisation before growth and flowering can occur. This is especially the case with fruit trees.We might not particularly enjoy a cold winter but our apple trees certainly do and produce a better crop in response to a lengthy cold snap.

In warmer climes many plants will require artificial freezing before flowering. Tulip and Hyacinth bulbs are examples. More importantly winter wheat seeds can be chilled before germination to enable a twice-yearly crop to be harvested.

Many vegetables can require in excess of 13 hours of daylight but may also require a cold period of vernalisation as well. Research is on-going into the climate change effect of rising temperatures on the important agricultural growing conditions involving vernalisation.

Of course, like all gardeners, a plant needs a period of rest – 'dormancy' when normal botanical activity ceases. The best example is winter dormancy in colder climates where growth will stop and not start again until the ground warms up and light levels increase, stimulating hormones into action. There are also seeds that require a little extra help before they are ready to leave dormancy behind and set out on the path to germination, be it mechanical cutting (scarification) or just a period of extreme hot or cold to kick-start those hormones.

It is to be noted that mechanical pruning will often stimulate new growth, but grafting may not work while a plant is in its dormant state.

Keen to propagate some new plants?

Vegetative (asexual) reproduction means growing a new plant from part of the existing one. Left alone plants will happily reproduce all by themselves as a single parent but a little human intervention makes all the difference.

Stolons or runners sprout from an existing stem. They sit on top or just below soil and produce long lateral stems. Plants with lateral stems can be propagated by layering, strawberries are a good example. As these surface roots grow they will produce new shoots, new strawberry plants and in time more strawberries – pass the cream please.

Rhododendrons are just one of many shrubs that can be air layered whilst still attached to the parent plant. I very much enjoyed doing this as a horticultural student.

Crocus Corm

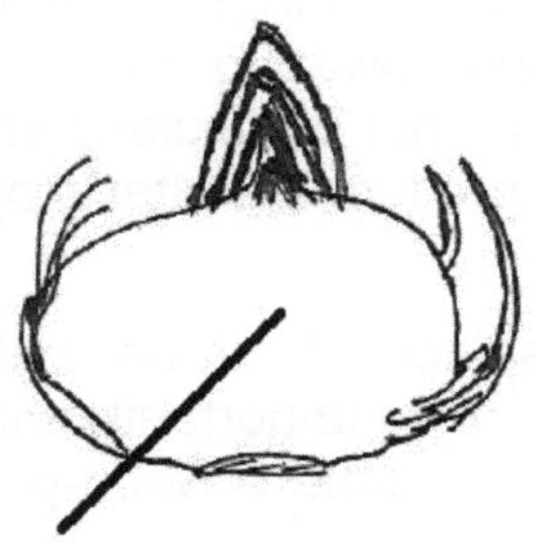

Solid Swollen Stem

Bulbs and corms propagate vegetatively underground, producing smaller versions of themselves. Carefully dig up any overcrowded clumps, detach the newly produced bulbs and corms from the parent and replant the healthy material.

Some bulbs, such as lilies, also produce bulbils on their stems, and these can be removed and grown on in seed compost to produce larger bulbs.

Rhizomes are underground stems which spread rapidly and produce many shoots along their journey which break cover and produce new plants. Rhizomes, Stem Tubers, Root Tubers – can all be propagated by cutting off a piece of the parent plant that is showing a new shoot or bud and replanting it. New plants from old!

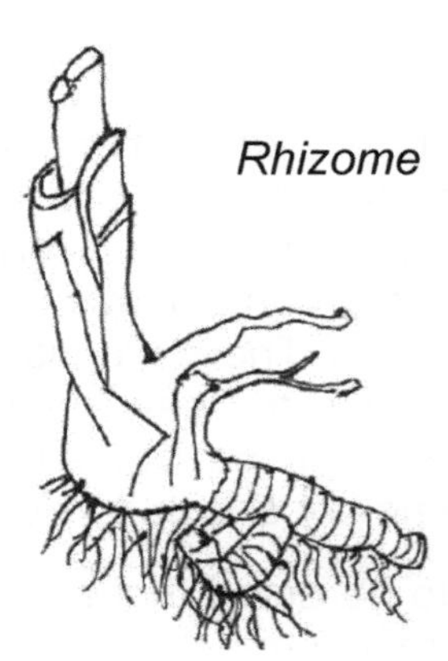

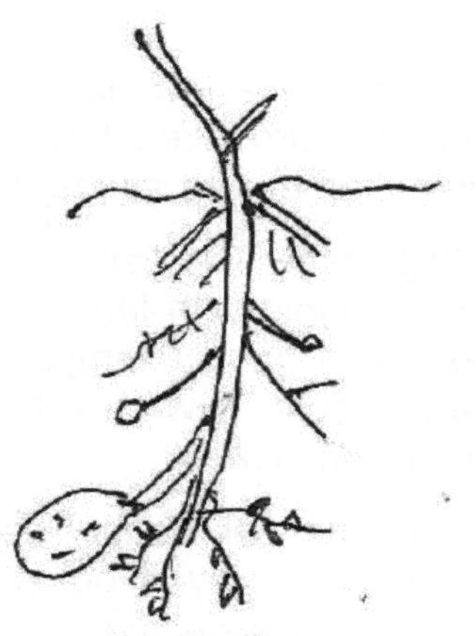

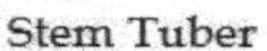

Stem Tuber

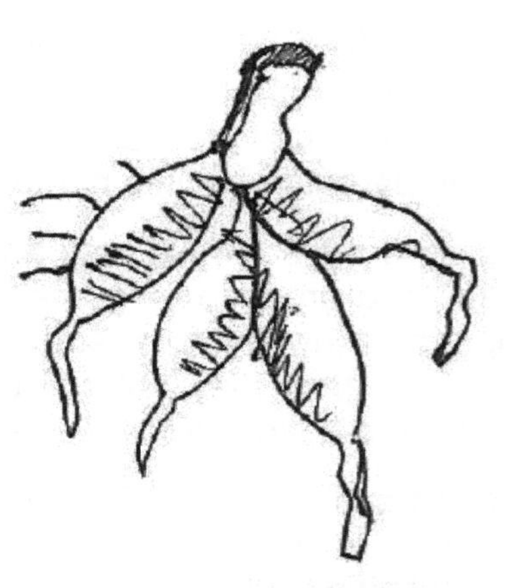

Root Tuber

Some plants will produce clone seeds by self-pollination (Apomixis) if they are unable to attract the opposite gender. *Rudbekia fulgida* 'Goldsturm is an example of a flower producing clone seed – every flower will be identical - useful for the floristry industry.

The transgender avocado can change sex from female to male in hours. During the day the flower opens making itself available for any interested pollinator, but overnight it closes and the following day opens with the male anthers at the ready. It is thought this rather strange behaviour has evolved to help in cross-pollination, and happily self-pollination is a less preferred occurrence in plants.

If you have an herbaceous perennial that has grown too big for its allotted space, dig up the whole plant. Prise the roots apart breaking it into two or three separate pieces replant them wherever you have an appropriate space to fill. So long as the new plants have a root system attached they should flourish.

Grafting is the technique of transplanting living tissue from one plant and inserting it into another, a tremendously important part of horticulture used particularly in the growing of fruit trees and also vines for wine production.

Many different fruit tree root stocks are available, number coded according to size.

The scion which is specially selected to contain the genes and the characteristics of the required fruit, apples, for example, is grafted onto the chosen root stock. The vascular systems of each plant cleverly join together. The root stock grows and the fruits borne have the same characteristics as the scion.

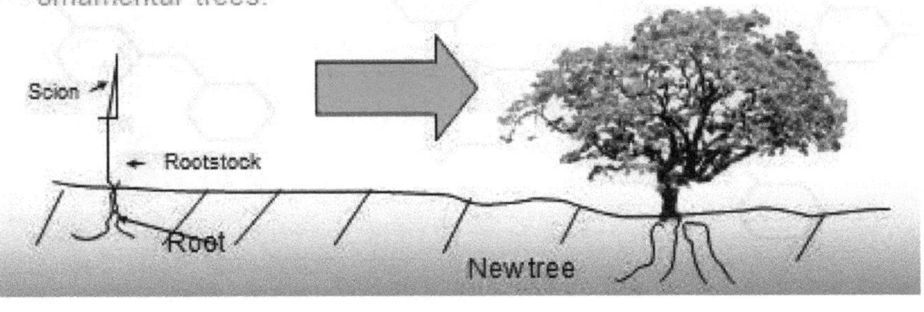

Using this method of propagation means that the eventual size of a tree can be controlled - we can't go scrumping if the tree is too tall! Growing from seed would take too long and rarely comes true - we need a constant supply of the same variety of apples and using a favourite scion ensures producers are growing a variety that they know will sell well.

In the case of wine, a different variety of vine can be grown simply by replacing one scion with another, leaving the growing rootstock in place. Tastes change – a fruity red for me please! Happily some time ago when disease destroyed many European wine vines, an American root stock was available to save the day. We really do need to continue with that special relationship!

I remember some years ago at Horticultural College seeing an amazing graft plant. It had a spectacular mix of flowers from the two grafted plants, yellow Laburnum and purple Broom, both of which hail from the same pea family - an excellent example of related species that have compatible vascular systems.

+ Laburnocytisus 'Adamii'
(a small tree in the Leguminaceae family)

A horticultural graft with a difference, called a chimaera – it is not a hybrid, it is a mixture of both parents Laburnum and Broom (Cytisus) and bears flowers of both parents.

Next I will deal with sexual reproduction (no giggling in the back row please)

Plants can reproduce by division –
humans of course cannot!

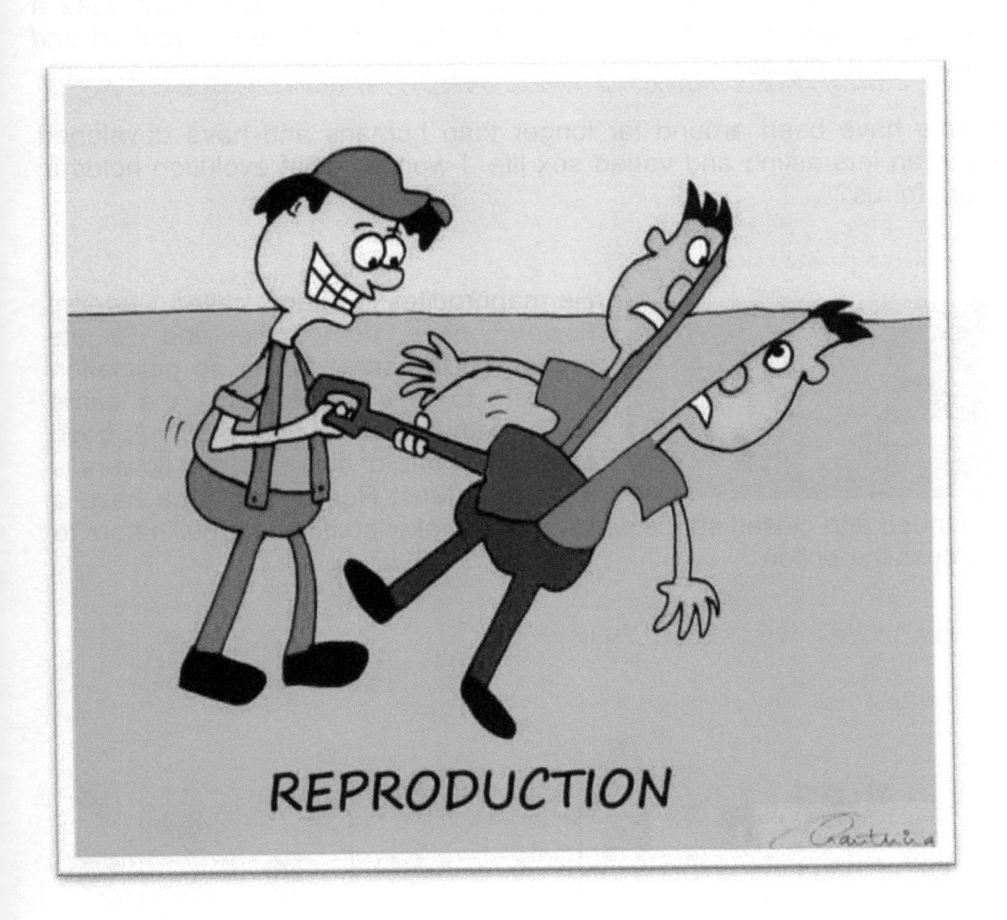

Chapter 9 Sexual Reproduction

OK - I'll admit it - before eventually retiring I spent a very happy 17 years teaching science to secondary school pupils. The response to sex education was always rather mixed between "Yuk is that really what happens" to "Who didn't know that sir!" Although I think one young lady might have been slightly confused. When asked in a test what does a baby take from its mother while in the womb she answered "yoghurt and mash" - but I digress.

Plants have been around far longer than humans and have developed quite an interesting and varied sex life. I wonder what evolution holds in store for us?

A perfect flower, the rose

Hermaphrodites, often called perfect flowers, have both male and female organs on the same flower so pollination is likely to take place within the same flower without any outside help. A good example of one of these perfect flowers is the Rose. Most Roses of course have a rounded and closed shape which would make it difficult for pollinators to access the pollen.

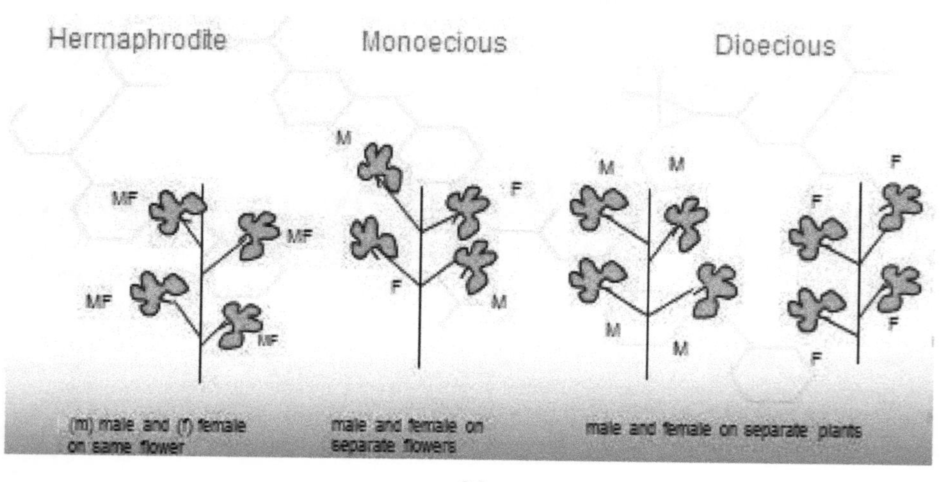

Monoecious plants also have both male and female organs but these are found on separate flowers found growing on the same plant. For example the Hazel (*Corylus avalana*). Monoecious plants are likely to be pollinated by wind as well as insects.

Dioecious plants still believe in segregation with flowers with male organs on one plant and those with female organs on another. Typical examples are the female fruit bearing berry plants such as *Skimmia, Ilex,* Laurel and Yew. These plants will require a nearby male plant if fertilisation is to take place and the production of berries to continue.

Help stop the sneezing! Female plants do not, of course, bear pollen and are very important in an allergen free planting scheme.

The sex cells of plants are referred to as gametes. The male sex cell is held in a grain of pollen, the female sex cell is an ovule held in the ovary. The male and female gametes need to get together – they need the help of a pollinator.

Do you remember those early Biology lessons when you had to draw and label the sex organs of a plant? It was a long time ago for me.

Stamen

The male organ is the stamen which consists of the filaments (stems) and the anthers which produce the pollen

Stigma

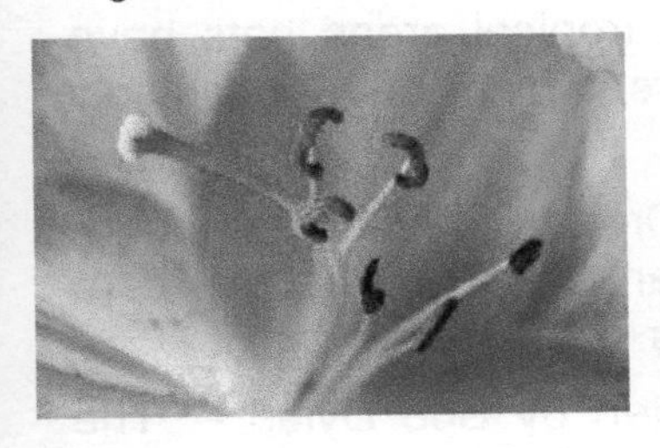

The female organ is the carpel which consists of the stigma the surface of which attracts the pollen and the style a tube joining the stigma to the ovary allowing the pollen to be carried down to the ovary.

Plants have adapted to attract pollinators, with their coloured petals, enticing fragrance and sweet tasting stores of nectar.

Such a happy sight in summer, watching bees and insects buzzing from one flower to another, transporting the often sticky, spiked or wrinkled pollen grains (all varying flower adaptations to help them cling on) as they are carried by the pollinator from the male anther to the waiting female stigma.

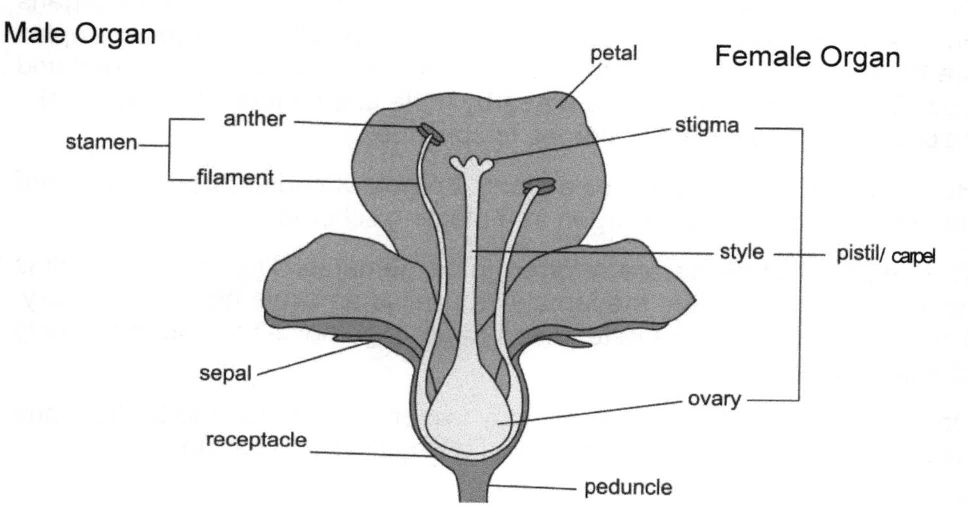

A longitudinal section of a simple flower

That's the bees – what about the birds I hear you ask. They play their part, searching for nectar and pollen grains which catch at short hairs around the base of the bird's bill and are then transferred to the next flower the bird visits.

Mice and other small rodents taking an early evening stroll around the garden can also help to transport pollen as can perhaps one creature we would not immediately think of – the bat. In tropical areas bats have become good pollinators, foraging in flowers which have obligingly adapted to open at night.

A good example is a tropical plant called the Organ Pipe Cactus, found only in the diverse and biologically rich Sonoran Desert in the Mexican Arizona region, which is pollinated solely by bats.

Who remembers that long ago song lyric written by Bob Dylan – 'The Answer is Blowin' in the Wind? Let me tell you, wind carries a lot more than answers, or even crisp packets and sweet wrappers – the wind carries the pollen of many important grass food crops, wheat rice, corn and barley

Wind pollinated flowers produce abundant amounts of small, light pollen making it easily carried by wind, although they may have sacrificed showy flowers and nectar stores. Male anthers tend to droop helping to shake out pollen in the lightest breeze, some are the typical tassel-like catkins. The female stigmas of the waiting plant are often cup shaped so that they can trap the wind-blown pollen.

Birch catkins

Usually found growing on long straight stalks, catkins are found typically on birch, oak and alder species.

Unsurprisingly, very few plants are pollinated by water. It seems rather a difficult process. Female parts of the flower form on long stalks reaching to the surface of the water. Strangely, the male flowers are released under water, float to the surface where they meet up with the female and transfer the pollen. As with wind pollinated plants large amounts of pollen are wasted.

Plants can be self-pollinating (monoecious) transferring pollen from the anther to the stigma of the same flower or a different flower on the same plant.

Other plants are cross pollinating (dioecious) transferring pollen from the anther of one flower to the stigma of a flower on another and separate plant. The most important thing to remember is that the second plant involved must be of the same or a closely related species.

Some monoecious trees such as the Common Oak and the Hazel produce separate male and female flowers on the same plant. This allows self-pollination among the flowers on the same tree whilst at the same time encouraging cross-pollination between adjacent trees as well. No wonder there are so many of them.

I have to say here that the flower knows a thing or two, able to identify pollen by its shape and chemical make-up, rejecting any that it feels may be from a somewhat dubious source such as the wrong species.

Pollen comes in many shapes and sizes

For instance, that pretty little anemone will not be accepting any pollen from that invasive bully, the giant hogweed.

Germination of a Pollen Grain

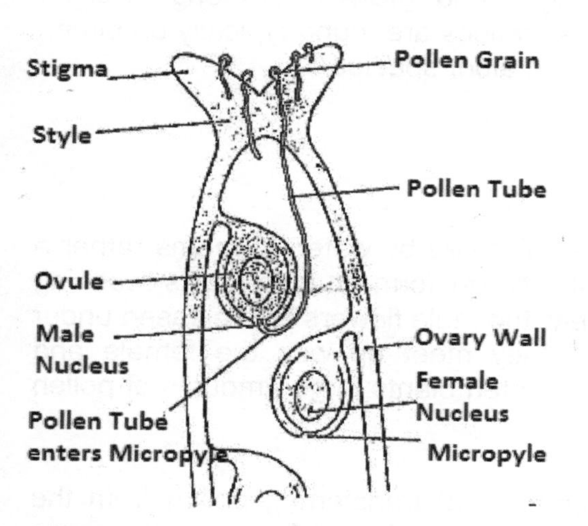

Germinated Pollen Grain on a Stigma

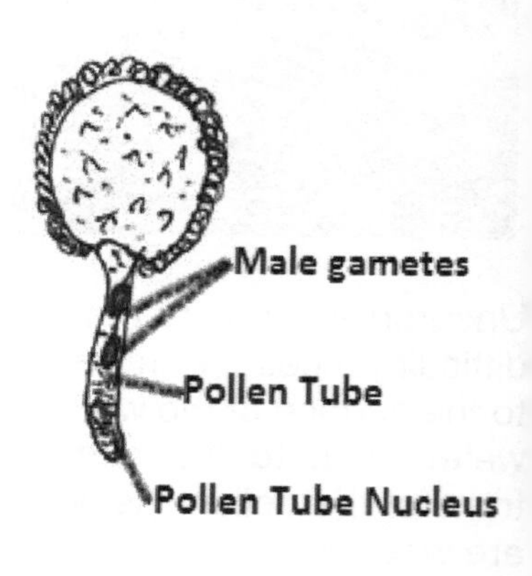

A pollen grain contains two nuclei; one forms a tube to create a pathway to the female ovule. The second splits as it journeys down that tube forming two further male sex cells.

Once inside the ovary, the first of these cells fuses with the female sex cell forming a fertilised embryo, which becomes the seed. The second cell has a vitally important part to play. It fuses with a special cell within the ovule and together they provide the endosperm which will act as a food store for the developing seed. Known as double fertilisation this only occurs with Angiosperms, the flowering plants. Clever stuff nature!

Much cell division then occurs, seeds are formed from the ovules held within the ovary which grows and develops a thick protective wall.

The pollinators have done their job - whether it be the birds, the bees or even the wind. The gametes have merged and exchanged chromosomes containing the genetic DNA contributed from each of the parent plants.

Let's take time out here to have a quick look and try to understand what is happening here – scientifically it is called genetics. Don't panic – keep reading – I'll try to keep it simple.

Genetics is the study of heredity, the characteristics we take from our parents. Plants are no exception. All plants differ in their make-up. These differences can be traced to the differences which are to be found in the genes, the basic units that determine the characteristics to be passed on to the next generation.

An Austrian monk, Gregor Mendel, made the first major discovery into the laws of inheritance in 1865 and so the study of the theory of heredity is often called Mendelism.

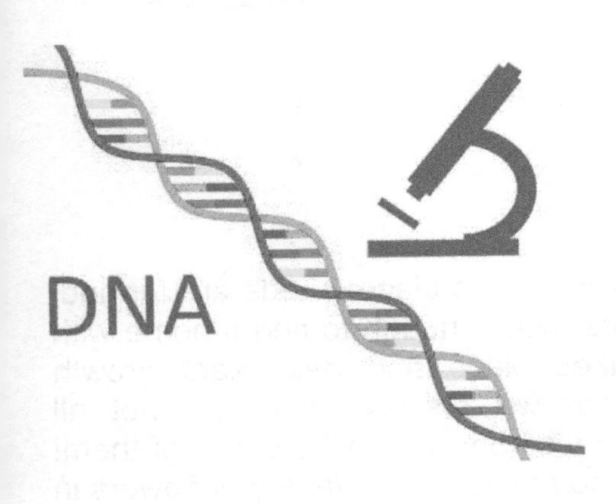

Chromosomes are part of the cell nucleus and usually occur in pairs, one inherited from the female and one from the male. They are long molecules of DNA containing the genes responsible for organising the biochemical life processes of each plant cell and for passing on the parental characteristics, for example, what colour the flower will be or how tall the grass or tree will grow.

When the male and female gametes (the sex cells) fuse at fertilisation a new combination of chromosome pairs (and therefore genes) are formed.

Different genes that govern the same characteristic are referred to as alleles. Alleles can be dominant, recessive or incompletely dominant. So when an embryo plant is formed if it has two matching alleles, for the characteristic of tallness, for example, it is said to be homozygous. If it has mismatched alleles, one for tall and one for short, it is said to be heterozygous.

A plant that has homozygous alleles will result in true breeding producing identical offspring.

The study of inheritance is detailed and complex and I will not cover it in further detail.

Epigeal Germination

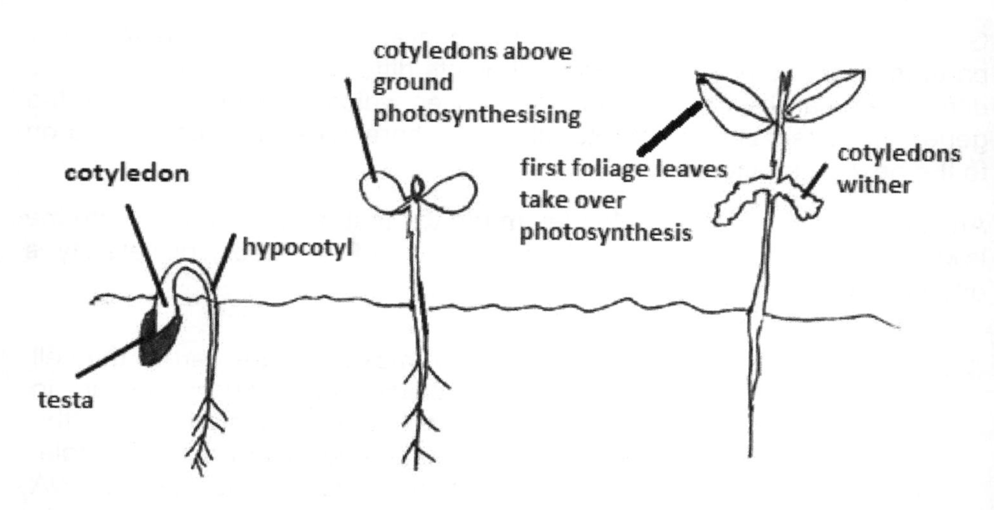

So, our flowers are pollinated, fertilisation takes place, seeds are formed, grow and are dispersed, and if they are lucky enough to find a home with the right conditions, germination takes place and new plant growth appears. Any new plants that appear will display some but not all characteristics of both parents but can well end up unlike either of them! It's called 'not coming true'. This can lead to a greater variety of flowers in terms of size, yield or disease resistance. The stronger plants will survive and those that cannot adapt or tolerate certain conditions will die out.

Developed from the ever expanding ovary, a fruit, which can be classed as either dry or succulent, surrounds and protects the seed necessary for new life until they are able to germinate.

Please do not confuse fruits with seeds. Seeds are the actual botanical fruit of the plant ovary whereas in a popular sense the fruit is the seed covering, not just something to be found at the greengrocer.

A simple fruit comes from one ovary which may contain one ovule (seed), for example a peach, or many, for example a tomato.

An aggregate fruit comes from more than one ovary contained within the same flower, a good example being those tasty summer raspberries.

Multiple fruits come from several ovaries within different flowers. As the fruit develops the ovaries fuse together. A pineapple is a multiple fruit.

Dry fruits have a hard covering known as the pericarp (fruit membrane wall) to protect their seeds and can be divided into three groups.

Firstly, they can be easily opened containers, such as pea pods or poppy capsules, which left alone will split releasing the seeds. Humans, of course, long ago discovered that if we harvested the peas pods before they split the contents would make a tasty addition to our dinner plates.

I've spent many happy hours at my mother's knee helping to shell the peas – whilst all the time looking out for the odd caterpillar that had found itself a cosy home. Much easier now it's all done for us and peas come frozen in plastic bags, but somehow that magic has disappeared.

These dry seeds which are ejected as the pericarp splits are known as dehiscent fruits.

Secondly, indehiscent fruits, nuts for example, will not split open of their own accord and need a squirrel with some sharp teeth or a human with nut crackers to release the tasty kernel within.

Thirdly, there is the combination dry fruit, with an outer case which splits open, releasing the seed which has a second and harder protective coating such as with the horse chestnut tree producing conkers.

Succulent or fleshy fruits on the other hand are often brightly coloured and mouth wateringly tasty. Oranges, the juicy flesh surrounding the seeds so delicious to eat, apricots and plums, damsons and kiwis. Tomatoes are also fruits as the seeds are contained within the fleshy surrounding, the definition of a true fruit.

Succulent fruits have evolved to attract animals (not just us humans) encouraging them to eat the tasty covering, exposing the seeds within, and so aiding dispersal and subsequent germination. So when that greedy blackbird devours all your cherries before they are actually ripe, or eats all your blackberries, leaving tell-tale purple patches behind, don't blame the

bird it is the fault of the tree or bush for producing such a tasty and colourful surrounding for its seed!

Fruit for dessert anyone

Seeds are often produced in large numbers to ensure the survival of the species and can remain dormant for many years even if harsh conditions such as long periods of rain, drought or fire kill off the parent plant.

To fulfil their purpose, seeds need to disperse, find a place to settle down and put down some roots. This will not always happen immediately. For example there are seeds which will lie dormant for a long period only germinating after a period of extreme heat such as that caused by Australian and Californian bush fires. Following these fires it has been found that certain seeds will germinate rapidly growing to replace vegetation destroyed by the fires. Clever nature again!

Some seeds make it difficult for themselves by requiring scarification (penetration of the outer seed covering) before germination can take place. This can involve a variety of procedures including scratching or treatment with acid and generally requiring mechanical, chemical or thermal help to speed up the germination process.

Others seeds require freezing and thawing to break dormancy, caused by those bossy hormones that inhibit germination.

The seed has a skin (testa) for protection, the density of which varies with the plant species. This skin needs to be broken down to allow the seed inside to take up water and so enable germination to take place. Small openings may need to be made to thick seed coats to allow water to penetrate. This can happen naturally where the seed coat can be broken down by fungi and bacteria in the soil, by the movement of soil or even by passing through a bird or animal in the first instance.

Annual plants produce some of the simplest seeds released from dry fruits when, following a nudge by those ever busy hormones, the pericarp splits allowing the seeds to germinate where they fall, allowing the continuation of the species.

Other plants such as grasses will wait until they are sufficiently dry before releasing their seeds which will often be dispersed by wind to seek a suitable place to germinate. Growing up between my paving slabs is not what I would call a suitable home, but somehow some grass always seems to end up there, along with a few weeds that I would prefer settled elsewhere.

For seeds to germinate they need to be viable, young and healthy in the first place and will need to find a home with the right conditions with access to plenty of water, oxygen (air) and usually warmth. Some seeds require light, others prefer darkness. A tip for gardeners is that very small seeds are best germinated on the surface of dedicated seed compost to get sufficient light.

If the seeds are lucky enough to find their perfect home, they start to take up water, begin to swell, the testa splits and germination can take place.

Germinated Seedlings

This process activates the important enzymes which help convert the starch stored in the seed into (glucose) sugar and proteins into amino acids or fats and oils. Sugar and amino acids build up roots, stems, leaves and buds.
The glucose is used for early respiration to provide energy for the seed store called the embryo.

First to appear from the embryo is the radicle shoot and this will always grow downwards into the soil to form a root which in turn will take up the water and minerals necessary for growth.

Next comes the hypocotyl shoot which will become the seed leaf or cotyledon designed to protect the emerging stem as the new seedling pushes up towards light.

Meristem cells swing into action and further roots develop and the stem elongates and breaks through the soil's surface.

Now the Cotyledons (the seed leaves), which are a seedling's next food source, are out in the open the stem straightens up and the first plant leaves appear. The process of photosynthesis can then take place and the cotyledons, no longer required, wither and die.

An emerging flowerbud

The first sign of that long-awaited flower is the sepal, a modified leaf, usually green, there to protect the flower bud, ensuring that it does not dry out.

As the bud gradually opens revealing the coloured petals within, the sepal remains at the base of the flower holding the petals in place. The petals surround and protect the male and female organs necessary to ensure the continuation of the species by pollination.

There are very special seed banks both in the UK and further afield where conditions are made perfect for each type of seed to be preserved for posterity. Other notable seed banks can be found in India and Colorado (Fort Collins).

Finally, I should mention here hybrids. A hybrid plant is produced by sexual reproduction between different species or genera.

Hybridisation is fairly common, and much easier, between different species, but very much less common between different genera.

When a Mahonia genera plant is crossed with a Berberis genera plant, both of the same Berberis Family (*Berberidaceae*), this combination produces an intergeneric hybrid named *x Maharberis.*

More likely and easier to cross to make a new species from two different species examples *Viburnum farreri* and *Viburnum grandiflorum* produces an interspecific hybrid named *Viburnum x bodnantense*, which I am sure is familiar to many as a fragrant winter flowering shrub.

Grafts, of course, are not produced by two parent reproduction, but by a specialist horticultural technique used to join two separate plants vegetatively, a sort of mixture of two plants.

There are now some amusing fruit, as well as vegetable, crosses, old familiar names being linked together to produce modern-day hybrids, and causing some amusement to audiences at talks I have given.

plum x apricot	=	pluot
tangerine x grapefruit	=	tangelo
cabbage x radish	=	rabbage
blackberry x raspberry	=	tayberry
lime x kumquat	=	limequat

I personally have not yet come across a rabbage in my local supermarket, although apparently this vegetable, which is leafy and takes on some of the colour and heat of a radish, was first trialled almost 100 years ago.

Chapter 10 What plant where – adapt or die

Like humans, plants have a favourite place to put down their roots. Well it wouldn't do if they all wanted to live in the same place.

In the British Isles and Northern Europe, very temperate and moisture rich regions, there are many garden plants which can settle happily in either a sunny or semi-shaded spot, dry or moist. Of course, there are always those plants that require something a little different to keep them happy.

Does your garden face north or south? Is it wet or is it dry?

For many gardeners in the northern hemisphere a southerly facing garden is ideal as it will maximise the sunshine hours available. North facing gardens will be much colder over winter months and may also not get sufficient warmth in summer.

Not all plants are happy in full sun, though. Some definitely prefer more shady conditions.

Some plants are happy with dry conditions and some will prefer to be damp, although not waterlogged.

Take time to look at your garden, see which areas have more sun than others, which more shade, and try to plant accordingly. A little time spent checking those hotspots (as well as those not so hot) will make all the difference.

Hydrangeas need plenty of water

Shady characters such as *Lamium* (Deadnettle) *Pulmonaria* (Lungwort) and *Hydrangeas* are happiest out of direct sunlight but do require a constant supply of moisture. When I moved to my new home there were several hydrangeas that had been planted at the front of the house in a south facing garden. They were struggling to cope with the heat and not looking their best. I moved them to a shadier site - much happier, my hydrangeas now flourish, but still make it known when they need extra water on a really hot day by starting to droop. Fortunately they recover amazingly quickly when given a bucketful of water, the contents of which are sucked rapidly up the stem by osmosis, replenishing the moisture required by the plant cells.

Water-loving plants, such as *Astilbe*, *Caltha palustris* (Marsh Marigold), Willow and Birch are all good beside ponds or streams very simply because they require a high water take up.

Camellias, to my mind are one of the most beautiful of the late winter/early spring flowering shrubs. They are happiest in a sheltered position and should be planted facing south or west as they will rapidly drop their buds if they become frosted in an east facing situation, caused not by cold, but by the rapid thawing brought about by bright early morning sunshine following a cold and frosty night.

Snowdrops brighten a shady area

Dry shade proves to be a difficult area for many plants but those with underground food storage organs, such as *Cyclamen* corms and *Galanthus* (Snowdrop) bulbs making their appearance in spring, happily grow under trees where soil conditions are notoriously dry. The storage organs contain sufficient levels of moisture and nutrients for survival and leaves bud earlier, maximising the available light levels ahead of the tree canopy opening.

Some plants like it hot! Roses give off their heady scent as the weather becomes increasingly warm. Some climbers, Ivy (*Hedera*) in particular, are happy growing up against a wall, having adapted to grow little root hairs to help them cling on, and some such as the vigorous *Clematis Montana* will continue to scramble aloft, snaking a path ever upwards making cutting back a somewhat perilous job.

It is always best to do a little research before planning a new look for your garden. A plant grown in the right place will be a happy plant, in the wrong place nothing will happen and an early demise may well be the outcome.

Sadly gardens these days seem to becoming ever smaller as we try to build as many new dwellings as possible on ever diminishing available land. Containers may be the solution here, but there are several obstacles that must be overcome if container plants are to thrive in the long term.

Simple, but effective

Choose an appropriately sized container – you will be surprised how quickly it can become overcrowded. Limited space for root growth could well be an issue if the pot is too small. Try to avoid plastic containers where the contents can become overheated or frozen. Water storage and retention is the main difficulty and artificial means may have to be put in place - I always like to add some of the very effective water gel crystals. Plastic bottles with drip feeders are also an option but, I find, detract from the beauty of the container and also, of course, require regular refilling. Moisture in a container is consumed rapidly so an established watering regime is an important factor in ensuring your container grown plants remain healthy.

Next, and I know it's obvious, but don't forget drainage. You don't want to drown the roots of your new plants where the main respiration takes place and a supply of air to the roots must be maintained. Make sure there is a drainage hole in the bottom of a container and add a few pieces of broken crockery at the base to help excess water escape.

Add the appropriate soil for your chosen plants – what pH is needed? In a container there will be no additional supply of soil or organic matter and no passing worms. The nutrients required by roots will be limited to those found in the growing medium selected so give them a good start in life by providing a well formulated product.

Don't forget if you are growing acid loving plants in a container use ericaceous (acid loving) soil and rainwater if at all possible, especially if your domestic water is alkaline and hard containing lots of calcium.

Rain water is acidic in varying degrees thanks to the carbon dioxide in the atmosphere producing carbonic acid that can cause havoc to forests. The UK ceased coal production for power stations that was decimating Scandinavian and European forests

It is recommended that the top couple of inches of soil in a container should be replaced on an annual basis, so introducing some new nutrients and organic matter. Eventually, of course, your plant may well outgrow the container space it has been allotted and you will need to provide it with a new home – sometimes easier said than done. If the plant has decided hat it does not want to be uprooted from a container, a surgeon's saw nay be required to cut into the root ball. I have successfully used this echnique on several occasions, but if it does not work for you, the only solution may be to break the container and plant afresh.

Sunflowers will grow, and then grow some more

A window box or several smaller pots could be an option to add colour to small spaces. Plant some annual seeds for a quick and colourful boost. Remember to keep all pots and containers well watered and cut flowers as they fade to encourage further growth.

I have always rather fancied a house with a sea view but there are of course problems to be overcome for the keen gardener.

Salt ions, minute charged particles in solution from the sea breeze, found in excess in coastal areas, cause severe stress to plants inhibiting the take up of moisture and nutrients from the soil through the xylem vessels. Consequently a plant can be significantly reduced in height and hormone production is disrupted. Too much salt can reverse osmosis, the flow of water up into the plant, instead moisture leaches out resulting in plasmolysis, the collapse of cells and the death of the plant.

Fear not, certain plants have adapted to cope with salty conditions by storing the salt, in due course releasing it through respiration. They also can grow extra succulent leaves which store water and release it to dilute the high concentrations of salt water. Brilliant adaptation!

Do some research if you are lucky enough to live by the sea, there are plants that will well tolerate these difficult conditions just look around and list them. If possible plant a hedge as a sea wind break and some vegetables in particular will appreciate the protection of a cloche. Raised beds where loam and compost can easily be replenished are also a good option.

It is possible for a human unhappy with his or her surroundings to move. As gardeners we can choose what we think is an appropriate place for a particular plant, and if it is not happy we can move it. An unhappy plant growing in the wild must adapt or die.

It has been a long and sometimes perilous journey, but over millions of years plants have evolved and adapted to live happily on this planet, thriving where surroundings suit them best.

It is thought that seeds from some of the very first *Cycad* plants, present in the first super continent, one giant land mass given the name Pangea, spread as tectonic plates shifted and continents drifted, taking up residence in various locations throughout the world as we now know it.

The reasoning behind this theory is that the huge and heavy Cycad seeds are denser than water and sink so must therefore have travelled along with the moving land masses – what a journey that must have been!

We know that plants have evolved over millions of years, most coping with massive and sometimes violent changes in climate, and adapting to survive in some of the more challenging growing environments found worldwide.

Moderate geographical climates have over time allowed much greater plant growth diversity. The greater the climatic difficulty the fewer plants that survive, but the cleverer they are in their adaptation.

More extremes of temperature will mean fewer plants, but those that do grow have adapted well.

When faced with adverse conditions, what must a plant do in order to survive?

First and foremost it needs to maximise oxygen extraction so it can breathe (respiration) and then seek out all available water to take up sufficient nutrients for growth enabling photosynthesis to then produce the necessary carbohydrate sugars glucose for growth.

Water is vitally important, of course, but the amount needed very much depends on the moisture retention of the growing medium, the soil. Is there a constant supply of regular rainfall or are there any other available sources of water such as from rivers, lakes or even underground.

A plant needs to protect itself from the elements, whether it be the scorching conditions of open prairie land, the freezing conditions of an alpine environment or the hot, wet and stifling conditions of a rain forest. Flooding is linked to high levels of deforestation.

Hungry predators must be deterred so that the plant has the best chance of growing to maturity and to ensuring the continuation of the species.

Finally, a plant needs a home, somewhere it can put down its roots, a location suitable for it to reproduce and continue the species. It needs to be anchored in soil or find a suitable location in water, or even seek out a friendly epiphytic host plant, willing to share its space in the world. It's a jungle out there!

Aerial roots

Tropical jungles can look like a mass of green and brown cabling that has grown in most directions hence the need for robust sharp machetes when exploring. I well remember these iconic Tarzan films! Did Tarzan bother to look down, I wonder, whilst swinging through the trees. So much greenery taking shelter from the scorching sunlight, having happily adapted to live in the dappled sunlight provided by the large tree canopies overhead.

Talking tropical, the grass plant Rice of the *Poaceae* Family really prefers high temperatures so long as there is plenty of available moisture, but has adapted to grow well in a controlled flooded environment, a safeguard introduced by man to protect this valuable plant from predators and ensuring the best chance of survival for this important crop.

Although happiest grown in regions of Asia where it originated, because rice is such an important crop it is now grown as short medium and long grain types in Mexico and the hotter southern states of the USA where agricultural progress allows seeds to be distributed from the air and crops strictly irrigated by computer controlled systems.

Interestingly in the USA and Canada there are hardiness designated zones, each with a 10 degree difference, which helps with selecting appropriate plants for hot summers and severely cold winters. Plants in the wild will already have selected and adapted to their preferred zones

I am constantly delighted when I look at different climate zones for species of plants especially trees, shrubs, agricultural food growing cereal grasses, non-cereals and vital vegetables. The selections have mostly been made by plants, even those specially introduced for food growing from other climates, it's still the plant that decides if and when to adapt and grow.

Some rubber plant species from the South American, Brazilian climate were tried out in South East Asia where their own native but different species rubber plants species grow very successfully. The results were disappointing; the plants decided they were not happy there!

Even in its native South American growing area the rubber plant can be tapped successfully in the wild but is severely hampered by its inability to adapt to leaf blight when grown in a cultivated situation, especially in plantations.

Elsewhere in the world plants have developed some ingenious ways of adapting to their particular surroundings

A prairie is a large area of level or slightly undulating land covered mainly by grass, but with some shrubs and herbaceous perennials and a few trees. It is a hot and dry environment in summer but winters are very cold. Prairie grassland soil is deep and rich in nutrients provided by the decaying roots of annuals, so there is a perfect soil combination of loam and organic matter.

A prairie landscape

Anything growing on prairie land needs deep roots to seek out water. Prairie vegetation tends to grow from underground the base rather than the tip which enables speedy re-growth following the many fires that occur as well as the decimation caused by hungry cattle chomping away.

Prairie grasses have narrow leaves to limit water loss and can bend easily to tolerate high winds. Trees and shrubs have found it much harder to adapt to the raging fires so few exist on prairies

Although we think of Alpine plants as belonging to a cold mountainous region, the definition of a true Alpine is a plant that grows above the tree line. Alpines tolerate severely low temperatures by growing close to the ground protecting them from extremes of snow and wind. Leaves are small to enable heat to reach the outer tips faster and roots grow deeper seeking warmth. Cells can dehydrate to avoid damage caused by freezing water. Mountainous Alpine plants will be subject to stronger ultra violet radiation with less dense atmosphere to absorb it and will grow a thicker protective membrane. Increasing amounts of sugar retained in the cells as well as the regulation of chemicals in stems and leaves all help to prevent tissue from freezing.

.Can you yodel?

There are Alpine plants that have adapted to live in a mountainous tropical situation where they have to cope with very high temperatures during the day followed by very cold nights. In Nepal and Northern India many Alpines sold as herbal medicines provide a sought after income, but this is putting the conservation of those plants under pressure.

Alpines can also be found happily growing at ground level on rockeries or in special planters, but do require a well-drained environment, they hate sitting in badly drained areas and just won't grow there

With a lack of pollinators to be found in extreme conditions, many Alpines revert to self-pollination, given the strange name 'apomixis', a process where the parent plant produces unfertilised seeds which are identical clones. The *Rudbeckia is* another plant that reproduces in this manner.

A second unusual process of vegetative reproduction has been given the name 'vivapery' where a very young embryo plant emerges but remains attached to the parent

Primroses are a good example of this sometimes called two-fold system of reproduction. Hardy alpine perennials, they have been cultivated to produce the many tough but multi coloured spring plants available to brighten our gardens after a long dull winter. Primroses can easily be propagated by digging up the plant following flowering and breaking it into two or three separate root pieces and then replanting. They will also produce seeds which spread far and wide. I have several tiny primrose plants growing along my gravel paths and I don't have the heart to remove them

Lichen

The Tundra, covering around 20% of the world's surface, has a particularly difficult and constantly cold climate with a great deal of exposed rock. There is little soil and few nutrients so any root system is of necessity shallow but a few hardy plants have adapted to survive under snow and to photosynthesise in extremely cold conditions. Leaves are small and flowers are ephemeral, produced quickly if and when temperatures rise

From one extreme to another, there are plants that have adapted to withstand scorching desert conditions. Desert plants send out roots deep underground seeking water which once taken up into the plant is protected by thick fleshy stems with a waxy coating. Succulent leaves become a fleshy store for water and in some plants the stomata, the little holes used to aid transpiration, close during the daylight as an aid to moisture retention. Photosynthesis subsequently occurs at night when it is cooler and the stomata holes reopen.

Feeling thirsty?

The desert cactus has been particularly clever growing spines to deter any animals desperately searching for a moist snack. Cactus leaves are succulent and vital for water storage.

Other typical desert plants are Agave, Aloe, many *Euphorbias* and the superb and strangely shaped water storing tree, the elephant tree, the *Pachycormus.* These trees withstand the most severe heat by dropping their leaves, vital photosynthesis continuing in the trunk and branches

Another weirdly shaped tree is the Madagascan Baobab of the *Malvaceae* (Mallow) Family of the genus *Adansonia* with a trunk that looks like a fat bottle, adapted of course for water storage, with a just a few small branches and leaves sprouting out from the top. Not just found in Madagascar, these trees also grow in Africa and Australia where severe drought conditions exist.

Where water is an issue, swamp and aquatic plants have adapted to survive under it so long as they can obtain sunlight for photosynthesis and minerals from the soil bed. Some have adapted to freshwater, others to sea water and some, such as plankton, happily float freely around the oceans. The oxygen required is often obtained by the adaptation of an air tube grown to the surface similar to the aquatic flower. Did you know that one of the earliest angiosperms was the waterlily?

Few gardeners will be able or wish to replicate the conditions found in a rainforest, but they are so interesting and diverse they do deserve a mention.

Rainforests are, of course, wet, but they are also humid and can be dark at lower levels due to dense overhead growth. They have just a shallow layer of fertile soil and plants send their roots sideways rather than downwards in a quest to capture nutrients. Trees grow ever upwards to reach the light, but shallow roots will not support a tall tree so many trees that grow in this environment have developed buttress roots (a bit like flying buttresses in big cathedrals) which develop on the surface all around the tree, supporting it and seeking out further nutrients.

Fan palms are a feature of the rainforest. They have large leaves good for catching sunlight for photosynthesis but which are segmented to allow any excess water to easily drain away.

Plants in a rainforest usually rub along well with each other. Lianas, woody vines at ground level scale the surrounding trees seeking light and then spread from tree to tree weaving an intricate network of stems which can reach up to 1,000 metres in length. They are over 2,500 species of these fast growing vines which ask nothing more than a leg up to daylight.

Epiphytes grow abundantly in this wet and humid environment including many flowering Angiosperms. They use a tree as a form of shelter, taking any moisture and nutrients required from the air, rain or any compost adhering to the tree's branches. Orchids are epiphytes well adapted to succeed, their large open roots able to take good advantage of the conditions around them. They produce hundreds of thousands of tiny seeds which are dispersed far and wide by wind .

Hemiepiphytes are not quite so welcome. Originating as epiphytic roots high up in a tree's canopy they grow slowly towards the ground. Once there they feast on the leaf litter available speeding up the growth process

The Strangler Fig

and then rather unkindly continue back up from whence they came becoming bigger and stronger and eventually strangling and killing the host tree. The Dreaded Strangler Fig (genus *Ficus*) is a good name for this species of ungrateful wild fig which belongs to the Mulberry Family.

The ecology of the rainforest is absolutely fascinating and an immensely important one in reducing the levels of unwanted carbon dioxide and influencing rainfall around the world. Mankind must urgently treat the rainforest with the respect it deserves and cease its destruction in our greed for timber, palm oil and soybeans along with the quest for further urbanisation of an ever increasing population.

Lonely!

We tend to forget one of the most difficult of growing media and that is volcanic soil. Although it is full of plant loving minerals plants cannot live on minerals alone – they need some soil and enough water to sustain them.

This happens very successfully in agricultural areas with volcanic history such as New Zealand, Indonesia, parts of Italy, the Canary Islands and Hawaii, many and varied locations where the soil is rich and fertile. Very few plants can grow in actual volcanic ash but in Mount St Helens in USA after the massive eruption that occurred destroying many animals and surrounding wildlife, plants were found to be growing back after a year or two with sufficient moisture and remaining soil contributing to promising fertile conditions.

Closer to home, some non-hardy plants grown in areas such as the temperate south-west of the UK, where temperatures are warmer and frost unlikely to be a prolonged problem, have adapted to be semi-hardy by increasing the levels of sugar retention in cells and will unexpectedly survive.

Chapter 11 Fighting the enemy

You have got the soil right, the positioning right, the planting just as you would want it. There is just one final problem to overcome– those unwanted garden pests and diseases – and, of course, the weeds!

First and foremost plants need to be cared for with good garden hygiene. At the end of what would seem a very long day of gardening, and believe me I really have to make myself do this, tools should be cleaned before they are put away. Firstly this helps prolong the life of the tool, but perhaps more important, it prevents the spread of weeds and pests from one place to another.

Examine your plants carefully for any signs of ill health. If in doubt take a photo and ask advice from the horticulturist at your local garden centre. The RHS also offers an advisory service for members which can be particularly helpful when you're stuck.

Plants can be fatally harmed by pests, diseases, and weeds. Prevention and control is required to reduce damage to a minimum.

On average there are around 40 main pests and a similar number of diseases, but don't let them put you off, learn how to deal with them.

Although insects may be the smallest of the pests you have to contend with, they are the largest class of animals on earth, approaching a million different species to date.

Fortunately only a relatively small proportion of these come under the heading of garden pests, but the list of those that do is still a long one, and beyond the scope of this book.

Encourage hedgehogs into your garden by leaving out a few tasty mealworms on a regular basis and they will come back for more. Hedgehogs have an amazing sense of smell so will sniff out any treats that might be available and even reward you by supplementing your generosity with a few slugs and snails on the side. As autumn approaches leave a cosy corner, not too well kept, and when Hetty hedgehog and friends are looking for their winter quarters to hibernate you may well find that they will accept your kind hospitality and decide to stay.

If you are attempting to attract a hedgehog or two, please abandon any use of chemically based slug pellets, the two do not sit happily together and may prove fatal to the hedgehog!

Keep hedgehogs safe

It is easy to go into your local garden centre where you will find many different chemically- based products on display aimed at eradicating a particular pest. Sadly many of these may have an adverse effect on wildlife as well as on those beneficial insects that could be attracted as a natural predator. Chemical sprays are not appropriate when young children are about and may release unwanted chemicals into the environment, causing damage to local ecosystems.

I am pleased to say that there is a growing supply of natural plant based non-chemical products that can be used to try to get rid of those pesky pests but best of all and perhaps rather surprisingly insects themselves are coming to the rescue in the guise of Biological Controls.

A few examples that have proved to work well:

Predatory mites (Phytoselius) attack their cousins, the red spider mite.

Predatory encarsia wasps will devour glasshouse white fly.

Eelworms (a microscopically sized worm-like parasitic creature) will attack vine weevil but unfortunately are also attracted to plants in the *Solanaceae* Family and will be very unwelcome visitors to potatoes and tomatoes which although quite different both belong to that family – so choose with care.

A friend in your garden

Mealy bugs will be attacked by a predatory beetle, the ladybird. These easily identified creatures (red with a varying number of black spots) are being used more and more by organic gardeners to help eliminate unwanted pests.

In fact a survey has found that there are 26 different species of recognisable ladybirds in the UK. The larvae of ladybirds, hoverflies and lacewings (if you can obtain them) are all partial to aphids and can be used in a greenhouse situation.

It's a bug eat bug world out there.

Enough for a whole family of thrushes

Invertebrates, those ghastly slugs and snails are gardeners' most unwanted and unloved pests, but hedgehogs and birds can be partial to them.

Nematodes are naturally occurring microscopic roundworms present in soil which infiltrate and infect their host causing it to die. The nematodes then feed off the decomposing body and reproduce. They are now being bred organically and environmentally friendly as the new ok to use slug and snail killer– ok for kids and pets, ok for wildlife, ok around crops but don't put them too near your pond, water snails are important.

Vertebrates such as deer and rabbits can be particular pests, munching their way through the choicest greenery without a 'by your leave', and then adding insult to injury by leaving you a calling card of their droppings for everyone to see

An unwelcome sight

There are a few loveable pests that we often encourage into our gardens. How many of us I wonder put out food for birds and squirrels, which may in turn attract the not so loveable rodents, rats and mice. Bank voles may chew some of your tender seedlings and who wants a mole hill in the middle of their well-manicured lawn.

A plant disease is said to have occurred by virtue of a change or malfunction in a plant's shape or size (morphology), its structure (anatomy) or its function (physiology). Basically we can tell when a plant is not feeling too well just by looking at it, but it is not so easy to identify the cause and put things right. To make matters worse there are non-biological as well as biological forms of plant disease.

I'll start with non-biological as these are probably the easiest to recognise and treat and are most certainly linked to environmental conditions.

First, of course, insufficient water but on the other hand, too much! A lack of nutrients brought about by poor fertilising. Damage caused by extremes of heat or cold, by hailstones, storms or frost. Damage caused by poor light. Damage caused by pesticides, by pollution or planting in soil with the wrong pH. These all come under the physiology heading and will affect the way the plant's system functions, in other words they will affect the growth of the plant. The various forms of biological diseases are more problematic and bacterial and fungal infections of plants and trees can be very serious

For better or worse, Fungus

Firstly, fungus, a gardener's bogeyman. There are around 100,000 different species of fungi, and around 10% of these may cause diseases in plants. Fungi are primitive organisms which have not yet evolved to manufacture their own food. They need to feed elsewhere so the majority take up residence on dead organic matter, while a small number are parasitic, choosing to attack living plants.

Dutch Elm disease was caused by a fungus imported on contaminated logs. The fungus was then spread through the trees by a beetle. This two pronged attack caused the death of over 80% of all UK native elms. Nature stepped in and many new saplings have re-grown from underground roots which were left behind when the diseased trees were removed. Trials have also taken place to produce a hybrid tree from surviving elm species which it is hoped will withstand the onslaught of both fungus and beetle. So far these have proved to be quite different to the UK native species and therefore not considered to be acceptable replacements.

Environmental agencies are now very vigilant and all trees and plants have to have a licence to come into or be taken out of the country, a plant passport

Fungi consists of microscopic strands called hyphae which join together to form mycelium. Reproduction takes the form of spores spread by wind or rain or vegetatively by the spread of mycelium across the ground or through soil.

With so many fungi reproducing at such a rate there are many forms of fungal disease. Well known is Honey Fungus (*Armillariamellea*) that will attack trees and shrubs such as apple, lilac and privet. Initially spring foliage will wilt and then turn yellow. As the mycelium spreads it interrupts the plant's water supply, causing death within a few weeks to a small shrub or several years later to a large established tree.

The fungal roots can spread up to 10 metres outwards, so Honey Fungus is easily passed on to a neighbouring tree and mycelium may be found under the bark of infected trees.

To control this disease it is recommended that the infected stump be removed and in severe cases a trench is sometimes dug to halt the progress of the fungal roots

Damping Off is a soil borne fungal disease found in the warm damp conditions of a glasshouse and causes the collapse of seedlings. Soil should be sterilised with steam and build-up of water on the soil's surface should be avoided. Spread seedlings so they have space to breathe and grow.

Black Spot is a water-borne fungal disease common where conditions are wet and warm. Remove diseased leaves which commonly show a darkening in appearance and burn them. This disease often found on roses will not affect the flower, but may in time damage the strength of the plant. Avoid overhead watering of roses as this will only spread the fungal spores.

Powdery mildew affects plants found in overly dry situations. As the name would suggest it is powdery in appearance and settles on the upper surface of leaves. Water any affected plants regularly and mulch to retain moisture

To avoid attracting fungal diseases as far as possible make sure that any new planting material you choose is healthy. Tidy up and lightly rake the ground before you plant and continue this good practice by removing any dead, dying or diseased material on a regular basis. It is vital to ensure that your new plant has its own space to allow the circulation of air. Water plants in the morning if at all possible.

Bacteria are minute organisms. There are far fewer species of bacteria than fungus and consequently there are far fewer plants affected by bacterial disease.

One important bacterial disease to mention is Fireblight (Erwiniaamylovora) which favours members of the *Rosaceae* Family. It first appeared in the UK in 1957. The bacteria, spread by insects as well as small drops of rain, causes branches to wilt and leaves rapidly take on a burnt appearance. A bacterial slime may be produced in humid weather which causes even more damage.

Severely infected plants should be removed and destroyed by burning.

Viral infections of plants are rarer but deadly. The virus enters the DNA of the host plant via the cell nucleus and spreads through the whole plant. Viral diseases cannot exist outside their host but may spread by contaminated tools. Remember to clean your equipment after use.

Leaf mosaic is the most common of the virus symptoms. Leaf distortion, colour streaking on flowers, fruit blemishing and stunting of plant growth can all be virus related.

Viruses would seem to be rather fond of salad crops. Leaf mosiac virus, particularly attacks tomato plants, causing mottling and yellow discolouration, followed by a shredding of the leaves. Cucumbers succumb to leaf curl becoming yellow and papery. This is often the cause of over-watering.

There are no biological or chemical viricides and contaminated plants must be burned and not composted under any circumstances

Weeds are, of course, plants that are not wanted in a particular place and are often the result of poor soil management, extremes of pH acidity or alkalinity and weather extremes. There are the everyday weeds which can be easily be eliminated by hand and then there are those weeds that are invasive, such as Himalayan Balsam and Japanese Knotweed which come under Legal Controls from the Environment Agencies of many countries

Ground Elder a perennial weed makes an unwelcome appearance in many gardens, spreading rapidly through a system of creeping rhizomes.

Dig down to remove as much of the root system as possible, not an easy task, and be prepared for it to return.

Nettles should be removed before they seed. If you do have space to keep just a few they are a particularly favourite habitat for butterflies.

Putting down a layer of mulch in late winter or early spring does help in suppressing weeds. I used a layer of manure compost in my garden at the end of last year and as well having good nutritious soil benefits I have noticed that thankfully there have been far fewer weeds emerging.

If you are using a spray weed killer be very careful where you aim as surrounding plants may not take kindly to being splashed and become unsightly and badly damaged.

To summarise, there are three main options we can use to counter pests, diseases and unwanted weeds.

Human (sometimes referred to as **Cultural** control) - removal of pests and weeds by hand. Try this on the red lily beetle. It's not quite as easy as it sounds, as they tend to drop off onto the soil as soon as they sense your approach! A soil drench works well I've found. Collect snails and dispose of them thoughtfully – not over the fence! Entice slugs into beer traps if you can spare a can!

Rotate crops – it helps to prevent a build-up of particular pests and diseases in one place. Pull out as many weeds as possible as soon as they appear and remove any dead or dying vegetation.

Biological - using natural predators such as parasitic wasps to feast on the aphids which are trying to feast on your plants. Nematodes are useful against slugs.

Chemical - there are many pesticide products available, designed to deal with severe infestations of pests and diseases. They should be treated with respect, used as little as possible, and care taken not to use them where they may be inadvertently taken by children, pets or wildlife such as hedgehogs so do try to avid use in domestic gardens. These chemicals can destroy food chains and do more damage than good in our gardens it is better to remove the infested plant as a last resort. Before taking such drastic action there is an ever increasing selection of 'organic' sprays or treatments available to try first.

Lastly, **Integrated Pest Management** is very simply the gardener's option of combining the best and most appropriate way of eliminating those dreaded pests from the choices available.

I would always recommend looking on-line to identify a particular pest or disease and then approaching your garden centre, club or even a local knowledgeable allotment holder to determine what organic or biological treatments may be available. It is far better to let nature work alongside you to maintain a healthy garden rather than you damaging the balance of nature with unwanted chemicals

Chapter 12 A growing debate

United Nations ... World Governments
Leading Horticultural and Science Research Bodies
all consider the problems of

Water Supply

ClimateChange

GM CropsResearch

OrganicGrowing

Biodegradation

Hydroponics

Plant Growth Research

Aeroponics

Deforestation

Soil Conservation Management

Science is the knowledge obtained from the study of the natural world through observation and experiment.

We owe an awful lot to all those scientists who have researched and set out their findings so we can leave less to chance in attempting to perfect our gardens. Horticultural advances are important to us as gardeners – agricultural advances are important world-wide.

Scientists are always looking for more, for the best way of doing things, to see how life can be improved. Sometimes we may not agree with their findings, we think we know best, but remember there are always two sides to a story, and it is good to listen to each side before you make your own decision.

With an ever increasing world population it is necessary for there to be an ever increasing supply of food. Pests and diseases ideally need to be eliminated, crop yields need to be maximised. There is a constant demand for more.

Genetic Engineering is a new bio technology which has a huge future. It can be used to engineer a heavier cropping fruit tree, a larger vegetable, a sweeter tasting berry, a variance in colour, shape or size. A process which can be engineered in a far shorter timescale than if it was to be done manually by cross pollination trials.

The idea behind genetic engineering is that a plant can have new features added to improve its genetic make-up.

For example a plant that has been found to be particularly disease resistant can have the gene responsible for that resistance removed and inserted into the DNA of another plant to help it develop a similar resistance to disease. Some vegetable plants such as soya beans can be genetically protected against weed killers.

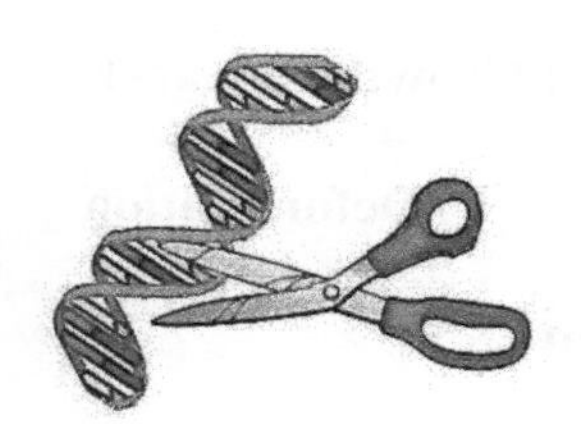

Genetic Modification adds a further meaning to the phrase 'cut and paste' where enzymes are used as genetic scissors. A further example is where special bacterial genes are introduced which are designed to produce toxins to target and eliminate certain pests without affecting the human food chain. The question is can we be sure that these bacterial toxins will remain safe for human consumption or will plants adapt and mutate naturally to destroy them and create different problems.

Another major concern is that GM plants may hybridise with their natural non-GM plant neighbours, starting a genetic chain of unwanted and unexpected changes in characteristics. Hybridising remember is a natural process of different plant species and genera exchanging genes and DNA to make new plants from seeds produced by two parents.

Only governments can grant licenses for GM experimentation. These tightly restrict the experimental growing area to avoid any cross hybridising with plants growing outside the control area. Breaches of the GM experimental licence have resulted in very heavy fines.

There is a very strong debate as to the morality of permanently altering plants for human advantage not fully knowing what the future biological consequences will be should unwanted and unpredicted mutations occur in the DNA of the genetically modified plant.

The science world is very split in opinion on the subject and a longer and thorough period of research is required before we can assess the effects on nature as a whole and whether GM crops will play an essential part in feeding the world or end up destroying those already successful crops grown today. I would like to believe that this research is not in vain and that eventually it will be possible to ensure that there are no future food shortages around the world. The big question is how long will the experimentation take and what will be the long term effect. Some research bodies say there is no evidence to demonstrate that any progress has been with GM crops – I hope that is not the case.

I have read arguments from both sides of the divide. One states that GM research is a distraction and all our efforts should be concentrated on looking after and improving our soils. The other says, that GM can only be a good thing and there is no science-based cause for concern and that GM crops have been grown extensively around the world for 20 years with no resultant health problems.

On the quiet I have also read that certain members of our Royal Family are at odds with each other over the question of GM crops; but we must all make our own choice

The natural approach

In the UK the John Innes Centre in Norwich is pioneering its own non-genetically modified strategy of producing the best possible crops.

Only the best will do

Research continues into plant structure and soil make up to find the best way of improving plant cell growth, crop yields, the take up of nutrients from the growing media and to see whether given the very best of conditions, a plant will naturally adapt and improve its growth pattern without the aid of genetic modification!

Germplasm is the term used for any genetically viable material that can be used to produce new plant and animal life. With plants the first thing that comes to mind of course is the seeds.

In fact, as long as the material contains cells with the necessary genes and chromosomes, any resource such as a leaf, a stem cutting, a root, even a pollen grain can be used to produce new plant life.

It is important to maintain stocks of viable material for research into future plant breeding or the preservation of a species in danger of extinction. These stocks may become increasingly valuable as genetic analysis techniques in botany advance.

As you would expect, this genetic material needs to be kept under strictly controlled conditions. Seeds, more easily perhaps in various seed banks around the world. Plant cuttings may be frozen and held until required, as with a human embryo.

The glasshouse at Kew

There are important international major seed storage units at the RBG Kew's Millennium Seed Bank at Wakehurst Place which hopes to conserve 25% of the world's species by 2020. This seed treasure house works withnearly100 countries around the world, some of whom duplicate the collection

An insurance policy for the future

Another major seed storage unit is the Svalbard Global Seed Vault, a secure seed bank on the remote Norwegian Island of Spitsbergen in the Arctic only 800 miles from the North Pole. Positioned over100 metres inside a sandstone mountain and the site of a disused coal mine the aim of this project is to hold a duplicate copy of every seed held worldwide. The seeds are heat sealed in special foil packets and then frozen. A pre-project study suggested that it was possible that the seeds would remain viable for many hundreds, if not thousands of years to come.

You may ask why it is necessary and is it worth the vast expense of maintaining such a facility. I believe it is just a case of being prepared, a case of 'what if". What if a certain plant was close to extinction? What if there were to be a major natural disaster or a nuclear catastrophe? What if a plague of pests devoured everything in their path for miles around? What if disease struck and wiped out those all important crops? It is good to know that in extreme circumstances seeds would be available to produce the crops necessary for mankind to survive

Nature's Mutation prompts Man's Manipulation

The science basis of GM growing is genetic mutation engineered by man. Hybridisation occurs naturally. A hybrid plant is produced by sexual reproduction between different (usually) species or (occasionally) genera.

A selectively bred hybrid is the result of pollination engineered by man (or woman) probably with a little brush taking pollen from one plant and introducing it to the other. The aim of this somewhat laborious process is to take the best characteristics of one plant and add them to the best of a second in an attempt to reach that all elusive perfection!

But - surprise, surprise who do you think got there first with genetic modification – nature of course!

Natural genetic modification or mutation occurs in many plants and is a change in the number of sets of chromosomes in each of the plant's cells. Research into this unusual phenomenon started in earnest almost a century ago when it was noticed that some plants developed an usually large flower, some a longer stem, and perhaps more commercially interesting a larger or juicier fruit! Some fruits were even seedless, which was puzzling, but easier to eat.

Time to call in the geneticists who quickly discovered that these interesting and unusual plant characteristics were caused by extra chromosomes being present – a mutation of each plant cell called ploidism.

Polyploidism kept simple. Most living cells that make up plant tissue and organs contain two sets of chromosomes, one set inherited from the male and one from the female. A lone set of chromosomes is called a haploid. Pair it with a set from the opposite sex and it becomes a diploid.

So far so good, but with a good percentage of plants something odd can happen.

This is where we arrive at a polyploid – a cell unusually containing more than two sets of chromosomes in its nucleus, the control panel responsible for directing all stages of a plant's development - caused by abnormal cell division during the reproduction process.

Polyploidy occurs naturally in up to 80% of plants. As the very first Angiosperms continued on their great evolutionary journey, there is evidence to show the on-going increase in chromosomes resulting from self-pollination as well as asexual vegetative reproduction.

Plants have far more chromosomes than humans

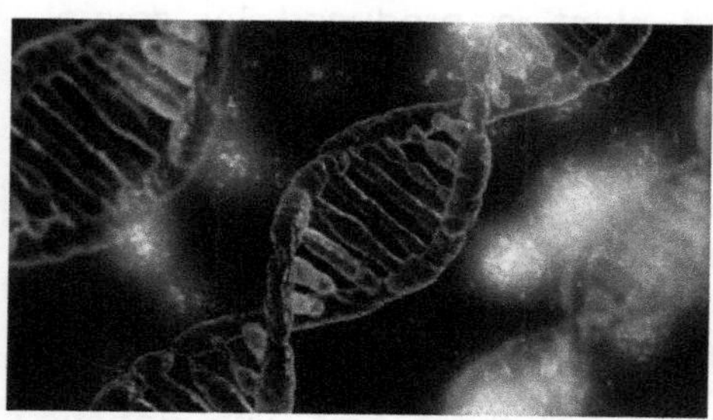

The increase in the number of sets of chromosomes contained in each plant cell will often lead to the plant becoming sterile. Certainly if there is an odd number of sets, the gametes will have difficulty in finding a partner in order to produce the seed necessary for survival.

So, not all modifications to a plant's growing pattern are caused by human intervention, but as ever we take what nature has provided and see how we can use it to our best advantage

Polyploid fruits tend to be larger overall, with bigger flowers producing bigger seeds in turn producing bigger fruits.

Genetic manipulation is used commercially to boost the size of fruit and vegetables by increasing the number of chromosome sets per cell, which is what polyploidy is all about.

Your melon's bigger than mine

It is not always easy. Water melons, with treble the standard number of chromosomes sets, are popular, but the cost of overcoming the sterility of large water melons by breeding with a normal diploid with just two sets per cell is quite difficult and expensive so production usually continues vegetatively. Sometimes it is easier just to let nature get on with the job!

A good example of nature's ability to mutate are the fruits, bananas, pineapples and figs, all of which will only grow vegetatively, without the need of an ovary at all. (I wonder why, perhaps their native climate wasn't favourable for pollinators when they first began their evolutionary journey?) Bananas, pineapples and figs all produce cloned offspring commonly known as pups (nothing to do with man's best friend!) which grow up alongside the mother plant.

The pros and cons of polyploidy. Plants that have mutated naturally may become larger, juicier, more disease resistant, but they may also become infertile because of mis-matched chromosomes.

I'm going to use grapes as an example. Simply because I like them and they are so much more enjoyable without pips!

No pips thanks originally to polyploidy

Seedless grapes were initially produced as a result of a natural genetic mutation increased chromosomes probably rendered one particular vine sterile. Although there were no fertilised seeds, the plant continued to develop the protective fruit from the expanding ovary, which became a seedless grape. A seedless grape, a bonus for the consumer, who much preferred not having to spit out a mouthful of pips.

A similar outcome can be achieved vegetatively without fertilisation of the plant ovules. In the case of grape vines this can be achieved by grafting or by artificially applying growth hormones such as colchicines or gibberellins to stimulate the plant's growth and halt fertilisation. Job done – no pips.

Plants that have been manipulated with hormone treatments become infertile because the growth of seeds has become restricted. In the example of the grape any small seeds that may remain are minute and lie dormant in the ovary – too small to be noticed when eaten.

So naturally polyploidy is a random occurrence but it can be artificially induced by the use of hormones.

Diploids have 2 sets of chromosomes per cell which is 'normal' for most plants and animals.

Tetraploids have 4 sets of chromosomes per cell. Examples: potatoes, rye, grapes, sugar, oats, buckwheat

Hexaploids have 6 sets of chromosomes per cell. Examples: bananas and apples

Octoploids have 8 sets of chromosomes per cell. Examples: strawberries, cotton, wheat, tobacco and kiwi fruit.

Polyploidy - not really simple at all, but an increasingly interesting subject.

An amazing man-made bridge, sits astride the Avon, home to an amazing product of nature, the Avon Gorge Sorbus tree.

This unique tree is a grand mixture of natural hybridisation followed by natural mutation, the extra chromosomes made it a polyploid but because there were an uneven number of chromosomes it became sterile.

Not yet ready to leave this world the plant reproduced by adapting even further to produce cloned seeds by a process called apomixis. One of nature's fighters and a fascinating case study of natural genetic plant diversity and adaptation being undertaken by Bristol University UK.

Soil and Water – Diminishing Resources?

We need to take soil management very seriously. The future of the planet's soil is of great concern with depletion of essential loam and nutrients in some areas reportedly approaching 20%-40%. Soil, a mix of loam (sand, silt and clay) organic matter, liquids, gases and billions of organisms all of which play an important role in supporting life on earth, took billions of years to create. It will not last for ever – we need to treat it kindly.

Recycling of nutrients both from small scale garden plots to vast agricultural practice is vital. What is taken out must be put back in. Soil will need to be topped up regularly with organic matter which will act as a slow release fertiliser and introducing more of those unpaid workers as well as more loam to replace lost minerals and nutrients. A healthy soil produces healthy plants and crops.

A compacted clay soil is not a healthy soil. It needs to be broken up to allow air and available water in, to let organisms breathe and roots to grow. A compacted soil is more likely to flood and the water will not penetrate but just sit on the surface or not be released.

Soil needs to be kept moist – too dry and it will become dust-like and be susceptible to wind erosion. It is better to have ground cover plants rather than leaving large bare patches of soil. The cover will provide habits for insects and worms, water retention will be improved and weeds suppressed.

Gardens and amenity space should be dug modestly, just enough to improve structure and air circulation, to make planting easier and more efficient.

Large scale agricultural ploughing adds to the global warming of the atmosphere by releasing huge amounts of stored carbon dioxide. World-wide trials are encouraged in an effort to reduce ploughing depths and are on-going.

Food for thought

World population is rising steadily. More people – more food - more soil - more WATER required.

People all around the world will need more water

We know that the world is not a level playing field when it comes to the availability of water. Some countries seem to have an everlasting supply; some suffer near permanent drought conditions.

What water we do have has been around for several billion years constantly being recycled. Very simply a cycle of evaporation into the atmosphere of water from the land, from vegetation, and in particular from trees which store enormous amounts of moisture.

Clouds in the atmosphere are there for a purpose. They collect moisture like a sponge, and when they can absorb no more, it condenses returning to earth as rainfall, draining into and refreshing existing water sources. The same amount of water to sustain an ever increasing population with ever increasing expectations.

Much of our planet is covered by ocean water, but only a small percentage of it is fresh.

Try to imagine living in those times when there was no running water, no sanitation, it doesn't bear thinking about but there are still places in the world where water poverty exists and water is not on tap.

We need water for agriculture; we use water in our gardens, parks and open spaces.

Designer buildings covered in plants, meant to be green, but how much water is required?

Land reclaimed from the sea.

Singapore has a superb new garden built on reclaimed land, but at what water recycling cost? The Aral Sea, which was one of the four largest lakes in the world covering a vast 26,000 square miles, has shrunk drastically in recent times as the rivers that fed it were diverted for other irrigation projects.

Water features and fountains, hoses and irrigation systems, showers, baths and fish farms. Humans sadly are not efficient in the use of this essential resource and this is becoming a major concern which needs to be faced by the whole world.

Don't waste this precious resource

I am not suggesting save water, bath with a friend, but just think twice as you turn on the hose, and think how lucky we are that we are able to do so.

The Organic Debate

Water and soil management go hand in hand we need both to exist.

Organic growing – is it for you?

It is possible that chemicals from sprays and an excessive use of fertilisers may affect wildlife and easily get into the food chain of animals. In wet weather run-off from fields can cause severe pollution to streams and rivers causing a situation where there is not sufficient oxygen dissolved in the water for aquatic organisms to survive – bad news for fish – bad news for any birds that rely on fish for their staple diet – bad news travels fast!

There is also a theory that chemicals used to spray crops may be the reason for the decline in the bee population which is a serious problem as we do rely on these busy pollinators. Although much research has being undertaken into the decline of bees and the use of neonicotinoid insecticides (similar in chemical make-up to nicotine) to date there is no conclusive evidence and consequently these products are not on the banned list

Naturally delicious

Organic gardening and crop growing seek to avoid the use of any manufactured fertilisers or pesticides, opting instead for a more natural approach by recycling nutrients from vegetative composting or well rotted animal manure and encouraging a more natural biological pest control.

This looks easy in theory, but in practice becomes very difficult for commercial crop growing, because of the vast amount of produce required. For the domestic gardener it is much easier to adapt to small scale organic gardening and something I myself am trying out.

Some dedicated organic farmers grow a sacrificial crop, such as the leguminous clover, known as green manure. Rather than being harvested the crop is dug back into the ground while still green returning nutrients, especially nitrogen, and improving soil structure.

Initially I would suggest that an organic approach is a more labour intensive way of gardening, but can be very rewarding if you are prepared to put in the time and effort required, and in the long run you could save money on expensive manufactured products, pesticides and fertilisers in particular.

First and foremost you need to prepare your soil well. Dig it over lightly to reduce compaction and allow in air, then spread some home-made compost, manure or leaf mould across the surface to add nutrients and help retain moisture. Worms will help you by taking this nutritious matter down into the soil.

Some organic gardeners prefer to adopt a 'no dig' policy for their soil for the main reason of allowing the biological systems in the soil to establish rather than being disturbed by the trowel and fork (secondly it saves a lot of hard work!) Large scale digging should be avoided as it can release stored carbon back into the atmosphere as well as disturbing the soil ecosystem being established. If you feel you can spare the space in your garden you need to make some compost, not always as easy as it sounds I have found, and be aware of what goes in, a compost heap can be a feeding ground for unwanted vermin if you try to compost meat or other protein waste.

Do not use weed-killers or pesticides in your garden, but take time out every day to check out what is actually growing. If any weeds dare to show their faces make it your daily mission to pull them up before they take hold, it doesn't take long if you catch them quick! Even larger infestations can be removed with a bit of digging, but do try to remove all roots and runners or they will return stronger!

It is possible to purchase effective biological controls to deal with pests, and I also like to encourage as much wildlife into my garden as possible. Birds, hedgehogs, toads and slowworms will all help to eliminate those unwanted slugs and snails. Far better to naturally suppress them than to resort to chemical based pellets and sprays which may well cause damage to the surrounding wildlife and insect ecosystem, particularly our declining population of hedgehogs, as well as to domestic pets.

If you are planting vegetable crops, look for those where the scientific research has been done for you and choose varieties that have been cultivated to be resistant to some of the more voracious pests. It is important to rotate crops on a seasonal basis to avoid a build-up of pests or disease associated with that particular crop.

When preparing the plot in an area with a large infestation of 'common weeds', rather than use a chemical weed-killer, try covering the area with a dark plastic or felt covering. The weeds will not grow without light, will rot down into the soil which will benefit from a good supply of chemical free nutrients especially from nettles – just make sure you wear some protective gloves or you will be looking for a dock leaf before you know it.

Mulching is vital, it helps to conserve water, improve soil and in very hot weather will serve as an extra protection for plant roots. Natural organisms love a mulch, warm and moist, just take a peep sometimes you will be surprised at what you unearth, lots of hard-working but unpaid worms and millions of beneficial organisms and microbes which of course you won't see.

Grass cuttings allowed to remain on the lawn give back valuable nitrogen into the grass roots especially in the heat of summer. Take the collector box off your mower occasionally, it will work well, just see for yourself. Better to recycle grass cuttings on the lawn, as too many in a compost bin can result in a somewhat slimy mess, which will only slow down the rotting process of the remaining compost

When slugs and snails are on the rampage, especially when ground and leaves are wet, I try to sneak out as dusk approaches with a pair of disposable gloves and an appropriate container and remove as many of the greedy pests as possible. An unpleasant job, I'm not sure that it makes that much difference in wet weather, but it makes me feel better!

If you are planting Hostas, choose varieties with thicker leaves, they are not so easy to for those hungry pests to eat!

Happily munching away

Companion planting can be successful in an organic garden. Grow mustard as a sacrificial plant with brassicas to tempt caterpillars away from your cabbages

Yellow nasturtiums attract aphids and whiteflies away from the plants you want to protect where hopefully they may be devoured by that ever hungry ladybird. One particular plant that seems to discourage a whole variety of pests: aphids, beetles, spidermites, ants and rabbits as well as discouraging black spot and mildew is garlic – I wonder why! Love the taste shame about the breath!

Garlic is also one of the strongest natural antiseptics. I didn't know that! An all-round amazing plant;' we should all find room for a bulb or two. Organic gardening is more time consuming, more labour intensive but highly recommended if it works for you. Studying nature's 'networking' is very interesting – it's surprising what we can learn and enjoy.

Climate Change

Changes in weather patterns are contributing to rain and hail storms which can cause severe damage to plants generally, but even more worryingly to food crops.

Soil can be eroded and there will be a loss of nutrients through flooding and consequent leaching away of the basic elements of nitrogen, phosphorous, potassium and calcium, not forgetting of course the minor nutrients - they all have an important part to play.

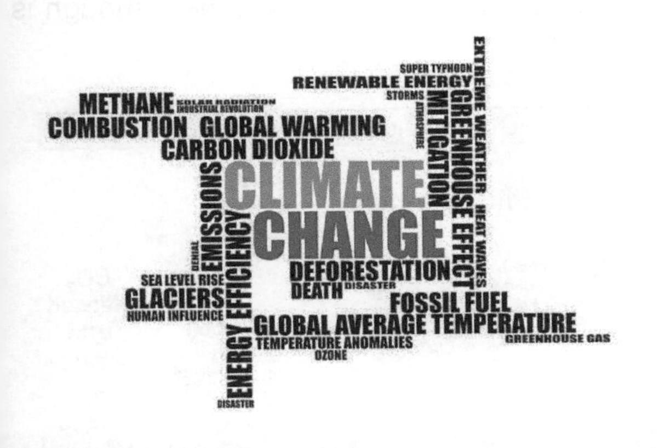

Flooding in south-west England, UK, in an area of Somerset called 'the Levels' is not unusual due to the geography of the area. One very extreme incident however lasted many weeks causing millions of pounds worth of damage to property and the surrounding land. Once the floods had finally abated farmers faced the huge task of decontamination and then replenishing the soil with fresh loam and considerable amounts or organic matter.

Hotter summers dry the land, and many of our native plants which are not used to such extreme conditions are finding life difficult. We may in the future have to look at different varieties of plants which are more sun tolerant.

Further problems are caused by periods of lower winter temperatures where many plants that may have been previously sufficiently hardy, will need extra protection or to be transferred to a more sheltered site.

Wind storm severity appears to have increased over the last 50 years and can cause obvious damage to plants; the great wind storms of the 1970s destroyed a huge percentage of trees in the UK.

Coastal wind storms will spread sea salt further inland which can damage certain plants and agricultural growing soils. Whatever the cause, evidence would suggest that our climate is changing and average temperatures are rising alarmingly at a faster rate every ten years

Getting world leaders to agree to treat global warming as a major issue is not going to be easy. Finding greener solutions to reduce carbon emissions perhaps even less so. What, if anything, can be done to stop or at least slow down the challenges which face us?

Where there is a will there is a way' as the saying goes. Where though is the horticultural will around the world?

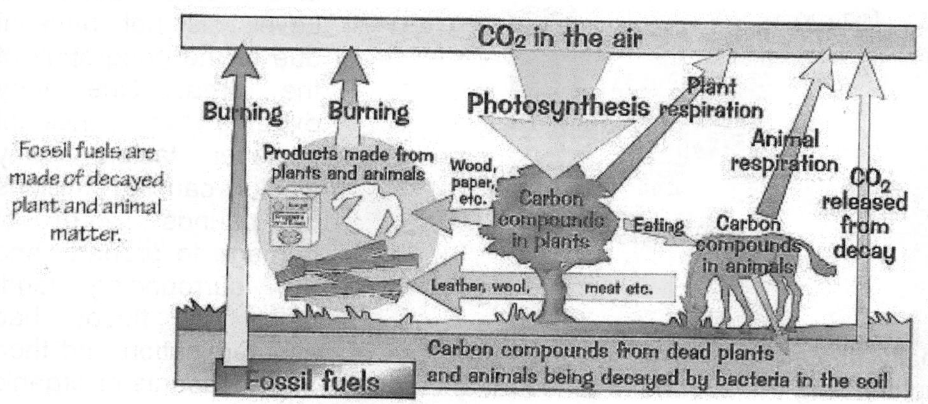

Carbon cycle (picture copyright CGP Books Ltd)

A problem with gas?

I wonder whether the world's gardeners and agriculturists can be persuaded to adopt a more organic approach. Less digging and ploughing means less release of carbons and methane into the atmosphere.

Recycling of plant vegetative matter, composting on a massive global scale, is required to refresh soil minerals and microbes for the nitrogen cycle.

Every little helps. As a science person I believe that global warming may stabilise by some kind of atmospheric equilibrium mechanism, but this can only be expected to happen if carbon omissions are severely controlled.

Biodegradation

An important area of research needs to be undertaken into producing biodegradable materials using plants.

It is painful for us to see many beaches around the world accumulating ugly piles of waste plastic products, but not as painful as it is for the wildlife that ingest this waste and suffer hugely as a consequence. A potential disaster ahead for us all and our wildlife.

The benefit of good, genuine and authorised biodegradable materials is that the break-down process can be relatively quick so long as there are available sources of light, water and, most important, bacteria. This is wonderful in theory, but in practice, even with good biodegradation small particles can still remain, even more dangerous than the larger ones!

There are stringent tests on any products that purport to be biodegradable as some only partially break down plastic, leaving smaller particles that sadly are even more easily taken up as a food source by unsuspecting wildlife.

A man-made disgrace

Chief UN scientists are looking for products that can easily be recycled, not just thrown into the sea in the hope that they will disappear.

Some said to be biodegradable products often contain rice or corn powder along with a plastic polymer with a short carbon chain such as polyethylene that should allow oxygen to attack the carbon chain and break it down to simple carbon oxides if the prevailing conditions are right. And that is the problem. There is much debate as to whether this type of degrading is beneficial. It is felt that it encourages us to throw away the product believing it will just rapidly degrade which is not the case.

Regulatory authorities internationally are cracking down on products that are not certified as degradable in composting or recycling facilities but instead requiring incineration with the resultant release of toxins into the air.

Plants have a huge future role to play in biodegradation. As a polymer science graduate I shall be following progress in this field with a more than casual interest in achieving a 100% biodegradable product.

Current research is testing one product made with the stems from spinach and parsley which can eventually be used in the manufacture of carrier bags, kitchen equipment and even computer components. That's interesting.

Bring back the paper bag

Whatever happened to good old paper bags from plant cellulose, easily made and easily recycled along with our paper waste? They may lack the strength of plastic but cause far less damage to our green and pleasant world.

Of course, it is not just plastic. In places not only our beaches but vast parts of our countryside are being contaminated. I know we are a throw-away society but just throwing away bottles, cans, even old mattresses and sofas is a disgrace.

It really is a case of 'Let's all try harder'. Let's all try harder to clean up behind us. Let's all try harder with recycling. Let's all try harder to make pollution a thing of the past. Let's all try harder to save our planet. Let's all try harder to make it a cleaner and safer environment for our children and animals.

No soil? No problem!

Experiments into planting without soil date back almost 100 years, when some research in the USA suggested that certain vegetable crops grown in a nutrient enriched water solution produced substantially larger crops in a greatly reduced timescale. Nutrients can either be organic or chemical.

Look no soil

Several methods of growing plants this way have been trialled with consequential developments of electronic monitors of light, lamps and cooling systems.

Plants are placed in the water solution together with a growing medium such as coir, perlite or rockwool which provide a support for emerging roots. Rockwool is not a natural product but is manufactured from chalk and rock, a process requiring very high temperatures and energy. It is not biodegradable, has dust and fibres which reportedly can prove harmful if inhaled and has a naturally high pH, which may not be the best for your plants .

The best way of growing plants hydroponically would appear to be in an open-ended trough through which a shallow film of nutrient rich solution is continuously pumped, and where roots grow to form a dense mat and provide their own natural support.

The conditions required for growth, i.e. temperature, light and nutrient levels and pH are computer monitored and controlled. Nutrient deficiencies are less likely to occur in a carefully regulated hydroponic nutrient supply system but careful monitoring and control is essential to keep these at optimum levels.

The energy saved by the plant in reaching down into soil for its nutrients is then used for growth, which can be up to 50% faster than when conventionally grown in soil.

It is of course very important that the circulating solution is maintained at a constant level that ensures that only part of the root system is covered. It is also essential to ensure that plant roots receive sufficient oxygen for respiration and that they do not 'drown'. This is economical in terms of the water and solution actually required, but perhaps not so economical in terms of investment in the computer equipment required to maintain the strict levels necessary to grow plants in this manner.

Another system of growing which has been trialled is aeroponics, first used by astronauts. Soil, pesticide and largely water free, seeds or seedlings are suspended mid-air in specially constructed chambers and misted with nutrient enriched water. The growing environment is clean and sterile, greatly reducing the chance of spreading the diseases and infections that often plague soil and other growing media. Water usage, as well as the amount of pesticides and fertilisers needed, are all greatly reduced.

Seedlings are happy with their roots hanging free and can be transplanted more easily avoiding stress and damage which may result when they are removed from soil.

It has been shown that tomatoes grown this way grow faster and produce a heavier crop. Is this the way of the future?

Plants in space

Maize seeds from the grass family were first launched into space in 1946 just after the Second World War. Scientists at the time were interested in the effect of radiation on plant tissue.

Sometime later a selection of seeds from conifer trees as well as seeds from the Sycamore and Liquidambar (Sweet Gum) broadleaf trees were flown around the moon on Apollo 14. No changes in growth were noted against other tree controls.

Overcoming the problems of lack of gravity, oxygen and light levels, astronauts have gone on to grow simple food crops at their space stations, some even being eaten - with no ill effect I am pleased to say.

In 1997 NASA sponsored in depth experiments aboard the Mir Space Station, choosing a high protein food crop of beans for their research. The beans grew faster than the control on Earth, which is interesting.

In 2016 school children were asked to help the **RHS** with their courageous **'Schools Rocket Science'** campaign. A large number of *Eruca sativa* seeds, commonly known as 'rocket' and a member of the *Brassica* Family, were launched into space hitching a ride in a Russian spacecraft.

The seeds spent six months in space and on their return with UK Astronaut, Sir Tim Peake, were distributed amongst the lucky children chosen who planted them and then compared the results with those less fortunate seeds which had remained earthbound. The children found that the seeds that had been into space, perhaps rather exhausted by their travels, took slightly longer to grow than those that were not able to go on the great adventure.

This experiment was valuable in raising the **UK schools STEM profile – Science, Technology, Engineering and Maths**. Who knows how many young children taking part in this experiment may go on to become the next generation of horticultural scientists.

Also last year a brightly coloured Zinnia was the first blossoming flower to be grown in space, proving 2016 to be a very good year for space research.

What is the point of this experimentation some might say? Is it likely that we will ever set up home on the moon or Mars?

What is the point of science? Science is the knowledge obtained from the study of the natural world through observation and experimentation and our lives depend on it.

I hope this simple and straightforward book has given gardeners of whatever age or experience an extra tool towards a deeper understanding of how plants with soil (or without) work together for the benefit of mankind. There is a lot to absorb, but like your gardens I hope your interest will continue to grow.

And finally ...

Chapter 13 The good news is...

And finally I can share with you something that I have known a long time; something I have often discussed with my wife and family, chatted about with friends and acquaintances, anyone who would listen really - now happily the time has come to report it as official - 'Gardening is good for you'.

I admit to there being times when I have overdone it. My knees letting me know they are there, my hands tingling from clutching the odd nettle, thorns finding a home where they have no right to be. All this pales into insignificance compared to the feeling of well-being that comes over me after I have spent a couple of hours in the garden.

Like many people I spend too long in front of a computer screen, could do with losing a little weight and do not sleep particularly well, but the fresh air time spent in my garden makes an enormous difference to how I feel, act and sleep.

It is not just the physical work of digging and hoeing, it is the mental 'switch-off' time that is all encompassing. Plants are not an argumentative bunch, they rarely answer back. Stress and worries vanish, replaced by a feeling of wonderment and joy as I discover a new shoot or an unexpected flower – all is not lost, there is hope of good things to come.

I am fortunate enough to have a front garden, not only does it give me great pleasure, there are many passers-by who stop to look at it and have a chat. Likewise I have enjoyed talking to others as I admire their gardening skills.

Gardens can become a talking point, something to break the ice, and perhaps brighten the day of someone who may through necessity spend many hours alone. A friendly chat is good for you, good for the passer-by and gives an enormous boost to that 'feel good factor'.

Try to spend a little time in your front garden, whatever the size. Even if it is the family car that is usually planted there, make room for a container or two, perhaps a hanging basket, or even a window box, it will bring a smile to your face when you return home and may lead to some interesting discussions with fellow gardeners.

There is growing evidence that not only physical but also mental health can benefit considerably from time spent outdoors in a garden.

Gardens designed for a hospital environment – somewhere to sit and gather our thoughts - to take comfort from. Gardens and open spaces where our children can play and let off steam and learn about the environment surrounding them.

Gardening is also having dramatic results in other areas such as prison reform, where inmates have found stability and focus, making prison gardens a good place to find peace, solace and time for reflection. In an increasingly busy, urbanised and industrial world, gardening can and does make all the difference.

A major meeting of the **RHS**, chaired by their Head of Science, agreed that there was overwhelming proof that plants – nature – horticulture - gardens – allotments have a very considerable impact on public health.

Officially, 'Gardening is good for you' – me, I just love it.

Tony Arnold

A growing friendship

Chartered Institute of
Horticulture

Enjoyed the book?

For monthly gardening tips, gardens to visit and topical news and views on the green world around us why not check out my website

http://www.scienceforthegardener.com

Keen to improve your garden? Dig out the Download.

Having difficulty in finding a speaker for your horticultural or gardening club? The Download is always available.

Teachers - seeking inspiration for your lesson plans? The Download may be the right answer.

Don't despair - help is just a few clicks away.

Use discount code DISC5 on my website to obtain a Download of 57 colourful slides and accompanying detailed notes for just £9.95 and see for yourself how a little science knowledge really can help to create a greener more successful garden.

Lightning Source UK Ltd.
Milton Keynes UK
UKHW02f1134050618
323742UK00007B/40/P